PgMP® Practice Test Questions

1000+ Practice Exam Questions for the PgMP® Examination

PgMP® Practice Test Questions

1000+ Practice Exam Questions for the PgMP® Examination

Dr. Ginger Levin, PMP, PgMP

CRC Press
Taylor & Francis Group
Boca Raton London New York

CRC Press is an imprint of the
Taylor & Francis Group, an **informa** business

Parts of *The Standard for Program Management*, 2017, are reprinted with permission of the Project Management Institute, Inc., Four Campus Boulevard, Newtown Square, Pennsylvania 19073-3299 U.S.A., a worldwide organization advancing the state of the art in project management.

"PgMP" is a certification mark of the Project Management Institute, Inc., which is registered in the United States and other nations.

CRC Press
Taylor & Francis Group
6000 Broken Sound Parkway NW, Suite 300
Boca Raton, FL 33487-2742

© 2019 by Taylor & Francis Group, LLC
CRC Press is an imprint of Taylor & Francis Group, an Informa business

No claim to original U.S. Government works

International Standard Book Number-13: 978-0-367-00113-1 (Paperback)

Visit the Taylor & Francis Web site at
http://www.taylorandfrancis.com

and the CRC Press Web site at
http://www.crcpress.com

Contents

Preface

When I was studying for the Program Management Professional (PgMP®) and was fortunate to become PgMP #2 in the world, I found in many cases, I was learning a new language. While I had a successful track record in managing programs – both large and small – I tended to think of my work in term of deliverables, not in terms of the benefits that would result. I also had different ideas about strategic alignment, stakeholders, and governance.

I found the first edition of the Project Management Institute's *The Standard for Program Management* to be illuminating such that I learned many approaches I had not considered but were useful ones I have applied since that time. It continues with the Standard now in its Fourth Edition.

In studying, I found a great way for me to best familiarize myself with the new terminology was to write down new terms, look them up, and personally think about how I would apply them to my work as a program manager. It led to development of a series of flashcards, which continues now with this edition. My hope is these flashcards will be useful to you as you become the next PgMP.

In studying for the PgMP exam, my advice is to read and think about the concepts in the Standard and then use this flashcard books we have developed. Then move on and test your knowledge with our *PgMP Exam Tests, Practice Tests and Simulated Exams* book. We hope these study aids will help you in your quest to become the next PgMP and wish you great success.

Of course, let me know your suggestions for continuous improvement and other ways I can help.

Dr. Ginger Levin, PMP, PgMP
Lighthouse Point, Florida

About the Author

Dr. Ginger Levin, PMP, PgMP, has more than 50 years' experience specializing in consulting and training in portfolio and program management, organizational project management, change management, and knowledge transfer. Dr. Levin had a career in the U.S. Government working in six agencies in transportation, including in the first agency-wide PMO at the Federal Aviation Administration and had a consulting firm in Washington, DC, focused on project management, maturity models and assessments, and organizational development. Since 1996, she works on her own in the project management field and is a PMI active volunteer. Dr. Levin also is an Adjunct Professor in project management for the University of Wisconsin-Platteville's MSPM program and at the SKEMA Business School in Lille, France in its doctoral program. She is the author, editor, or co-author of 21 books and has a book series with Taylor & Francis. She works with UT-Dallas in its PgMP and PfMP boot camps. In 2014, she won PMI's Eric Jenett Award for her contributions to the field. She received her doctoral degree from The George Washington University and won the outstanding dissertation award.

Introduction and Program Management Performance Domains

Questions

1. Describe the guidance from the *Standard for Program Management Fourth Edition*

2. What are principles of program management?

3. What does generally recognized mean?

4. What is good practice?

5. How are good practices measured?

6. Who determines what constitutes good practice in an organization?

Answers

1. The guidance is on principles, practices, and activities on program management that are generally recognized to support effective program management practices that apply to most programs most of the time.

---

2. Tenets held to be true and important for effective management of programs.

---

3. General consensus that the principles, knowledge, and practices are valuable and useful.

---

4. General agreement that the principles, knowledge, and practices improve program management and the chance of program success.

---

5. By the extent and effectiveness of benefits delivery and realization.

---

6. Organizational leaders, program managers, program teams, and possibly a Program Management Office (PMO) based on specific requirements of the program or its sponsoring organization.

Questions

7. What are other terms used for the program steering committee?

_____

8. Why is the purpose of PMI's *Code of Ethics and Professional Conduct*?

_____

9. What does the *Code of Ethics and Professional Conduct* require?

_____

10. What is a program?

_____

11. How are program components related?

_____

12. List three reasons when programs are better managed as portfolios?

_____

13. What is a component?

_____

14. What is the purpose of a project in a program?

_____

Answers

7. Program or portfolio governance board.

---

8. It specifies obligations of responsibility, respect, fairness, and honesty for use by program managers as they do their work.

---

9. Practitioners demonstrate a commitment to ethical and professional conduct, which means they carry with it an obligation to comply with laws, regulations, and organizational and professional policies.

---

10. Related projects, subsidiary programs, and program activities managed in a coordinated way to obtain benefits not available from managing them individually.

---

11. Through their pursuit of complementary goals that each one contributes to benefit delivery.

---

12. 1. If the components do not advance complementary goals.
    2. Are not jointly contributing to the delivery of benefits.
    3. Are related only by the common services of support, technology, or stakeholders.

---

13. Projects, subsidiary programs, or other related activities to support a program.

---

14. To generate outputs or outcomes required by programs within defined constraints.

Questions

15. What are five examples of defined constraints?

_____

16. What is another name for a subsidiary program?

_____

17. What is a subsidiary program?

_____

18. Provide an example of a subsidiary program.

_____

19. What are other program-related activities?

_____

20. List six examples of program-related activities

_____

Answers

15. 1. Budget
    2. Time
    3. Specifications
    4. Scope
    5. Quality

---

16. Subprogram

---

17. Programs sponsored and conducted to pursue a subset of goals important to the primary program.

---

18. If the program involved developing a new electrical car, subsidiary programs may involve developing the motor, battery, and charging technologies.

---

19. Work processes or activities designed to support a program but are not tied to subsidiary programs and projects sponsored or conducted as a program.

---

20. 1. Training
    2. Planning
    3. Program-level control
    4. Reporting
    5. Accounting
    6. Administration

Questions

21. What are two other terms for other program-level activities?

_____

22. What does the term activities mean in program management?

_____

23. What is the primary value of managing an initiative as a program?

_____

24. Why can program components be managed in an iterative but non-sequential manner?

_____

25. What are two ways in which a program can be initiated?

_____

26. What are four examples of a program initiated to support new strategic goals and objectives?

_____

27. What are three examples of a program formed from existing projects, programs, and other work?

_____

Answers

21. 1. Operational activities
    2. Maintenance functions

---

22. Program activities designed to support a program. It is not activities performed during the course of the program's projects.

---

23. The acknowledgement of the program manager's readiness to adapt strategies to optimize delivery of benefits to the organization.

---

24. Because of the need to adapt to the outcomes of the components and the potential need to modify its strategy or plans.

---

25. 1. To support new strategic goals and objectives.
    2. When the organization recognizes that its ongoing projects, program, and other work are related by their pursuit of common outcomes.

---

26. 1. A portfolio-based decision to develop a new product or service or expand into new markets.
    2. To influence human behavior.
    3. To ensure compliance with a new regulation.
    4. To respond to a crisis.

---

27. 1. A process improvement program by previously independent software development initiatives.
    2. A neighborhood revitalization program.
    3. A community outreach program.

Questions

28. What is program management?

_____

29. What is involved in program management?

_____

30. Who performs program management?

_____

31. What is the purpose of the five program management domains?

_____

32. What are nine actions related to the interdependencies of the domains to determine the optimal approach to managing program components?

_____

33. How does organizational project management interface with program management?

_____

Answers

28. The application of knowledge, skills, and principles to a program to achieve the program's objectives and obtain benefits and control not available if the program's components were managed in a standalone fashion.

29. Aligning components to ensure program goals are achieved and program benefits are delivered optimally.

30. The program manager who is authorized to lead the teams responsible for achieving the program's goals and objectives.

31. They are groupings of related areas of activities and functions that uniquely characterize and differentiate the activities found in one domain from the others in program management work.

32. 1. Define how outputs and outcomes contribute to the program's benefit delivery.
    2. Monitor benefit realization and ensure benefits remain aligned to organizational goals.
    3. Ensure outputs and outcomes are communicated so the program can optimize the pursuit of its intended benefits and provide value.
    4. Lead and coordinate program activities across components, work, or phases.
    5. Proactively assess and respond to risks spanning program components.
    6. Communicate with and support stakeholders for an integrated perspective of all activities being pursued in the program.
    7. Align program efforts with organizational strategy and the business case.
    8. Resolve scope, cost, schedule, resource, quality, and risk issues through shared governance.
    9. Tailor program management activities, processes, and interfaces because of cultural, socioeconomic, political, and environmental differences.

33. It is a framework in which projects, programs, and portfolios are integrated to achieve strategic objectives.

Questions

34. How do programs deal with change?

_____

35. How is management handled on a program?

_____

36. How is program success measured?

_____

37. How would you clarify the relationship between program management and portfolio management?

_____

38. List three ways as to how organizational strategies and priorities established as part of portfolio management support program management?

_____

39. How do the program and portfolio management functions support the organization?

_____

40. How would you clarify the relationship between program management and project management?

_____

41. How do project managers support the program?

_____

Answers

34. By accepting and adapting to it to optimize the delivery of benefits.

35. Program managers ensure program benefits are delivered as expected as they coordinate activities of the components.

36. By the program's ability to deliver its intended benefits to the organization and the effectiveness and efficiency in doing so.

37. Collaborative – both portfolio and program managers work together to ensure desired benefits are effectively and efficiently delivered.

38. 1. Defining programs to be pursued.
    2. Endorsing program strategies for delivering benefits.
    3. Allocating resources programs require.

39. By defining how the organization's strategic plan will be supported and delivered through prioritized and resourced programs.

40. Collaborative – program and project managers work together to define strategies to pursue program goals and deliver benefits.

41. By delivering outputs and outcomes that reconfirm or adapt the program's strategic direction and its components.

Questions

42. List two ways the interactions and relationships between program and project managers probably will change over the course of the program.

_____

43. List four roles of the program manager as the projects progress?

_____

44. How would you describe the interactions between the program and project functions?

_____

45. Why can a program manager influence a component manager?

_____

46. What are four reasons why the relationship between the program manager and operations is critical?

_____

47. What is the definition of business value?

_____

Answers

42. 1. In the early stages, the program manager may need to work closely with the project managers.
    2. During the work execution and closing phase, program managers focus more on coordinating interdependencies between projects; the project managers are focusing on managing project activities.

---

43. 1. Identifying and controlling the interdependencies between projects.
    2. Monitoring project performance.
    3. Addressing escalated issues from project managers.
    4. Tracking component contributions to the successful delivery of consolidated benefits.

---

44. Iterative and cyclical.

---

45. Because of the iteration and exchange of information and alignment of actions between them.

---

46. 1. The program impacts the operational activities or lines of business.
    2. The program's benefits may influence the scope of operational activities.
    3. Program deliverables may be transferred to operations to ensure benefit delivery is sustained.
    4. Both program and operations managers are responsible to ensure a balance between them and to support successful execution of the organization's strategic objectives.

---

47. The sum of the tangible and intangible elements of a business.

Questions

48. What are seven examples of tangible elements?

_____

49. What are eight examples of intangible elements?

_____

50. How can new business value be generated?

_____

51. In terms of generating business value, what is the role of program management?

_____

52. What is the key role of the program manager?

_____

Answers

48. 1. Monetary assets
    2. Facilities
    3. Fixtures
    4. Equity
    5. Tools
    6. Market share
    7. Utility

---

49. 1. Goodwill
    2. Brand recognition
    3. Public benefit
    4. Trademarks
    5. Compliance
    6. Reputation
    7. Strategic alignment
    8. Capabilities

---

50. Through portfolio, program, and project management to pursue new business strategies consistent with its future vision and mission.

---

51. To optimize the management of related component projects.

---

52. To lead the teams responsible for achieving program objectives.

Questions

53. What are seven examples of the program manager's role in managing and coordinating complex issues from uncertainties that may occur?

_____

54. What are nine examples of what program managers are expected to do?

_____

55. Since the program manager must balance the needs of the components and be able to adjust strategy or plans, he or she must be:

_____

56. List the six key skills program managers require.

_____

Answers

53. 1. Outcomes

    2. Operations

    3. Organizational strategies

    4. Resources

    5. External environment

    6. Organizational governance systems

    7. Stakeholder expectations and motivations

---

54. 1. Work within the performance domains.

    2. Interact with project and other program managers.

    3. Interact with portfolio managers.

    4. Coordinate with governance boards, sponsors, and the PMO.

    5. Interact with operational managers.

    6. Ensue the importance of each component is recognized and understood.

    7. Ensure the program structure and processes are effective.

    8. Integrate component deliverables, outcomes, and benefits into the program.

    9. Provide effective leadership to the program team.

---

55. Competent

---

56. 1. Communications

    2. Stakeholder engagement

    3. Change management

    4. Leadership

    5. Analytical

    6. Integration

Questions

57. Since the program manager must align the program's approach with the organization's strategy, deliver program benefits, collaborate with stakeholders, and manage the program life cycle, the program manager exhibits six competencies – list them.

58. What is the role of the program sponsor?

59. List two ways in which the sponsor provides guidance to the program manager.

60. What is the role of the program management office?

61. Are PMOs established only for individual programs?

Answers

57. 1. Manage details and take a holistic, benefits-focused view.

    2. Have a working knowledge of principles, practices, tools and techniques of portfolio, program, and project management.

    3. Interact seamlessly and collaboratively with the Governance Board and other executive stakeholders.

    4. Establish a productive and conducive environment with the team and their stakeholders.

    5. Leverage business knowledge, skills, and experience.

    6. Facilitate understanding and agreement.

---

58. A person or group who provides resources and support for the program and is accountable for its success.

---

59. 1. Ensures the program receives high-level support and attention.

    2. Keeps the program manager informed of any changes that may affect the program.

---

60. Standardizes program governance processes and facilitates sharing of resources, methodologies, tools and techniques.

---

61. No- they may be established for an individual program, but they may provide support to one or more programs under way in the organization.

Questions

62. List 11 ways PMOs can support the program manager.

_____

63. On a large or complex program, list three other areas of support from the PMO.

_____

64. What happens if the organization does not have a PMO?

_____

65. What is a program management performance domain?

_____

66. When do program managers carry out work in the domains?

_____

67. What are the five performance domains?

_____

Answers

62. 1. Define standard processes and procedures to follow.
    2. Provide training in these processes.
    3. Support program communications.
    4. Support program and change management activities.
    5. Conduct program performance analysis.
    6. Support management of the budget and schedule.
    7. Define general quality standards.
    8. Support effective resource management.
    9. Support reports to leadership and program steering committees.
    10. Support document and knowledge transfer.
    11. Provide centralized support for managing changes and tracking risks, issues, and decisions.

63. 1. Personnel and other resources
    2. Contracts and procurements
    3. Legal and legislative issues

64. The program manager performs these functions.

65. A group of related areas of activities or functions that characterize activities found in one domain from others.

66. During all program management functions.

67. 1. Program Strategy Alignment
    2. Program Benefits Management
    3. Program Stakeholder Engagement
    4. Program Governance
    5. Program Life Cycle Management

Questions

68. What determines the extent of activity in a domain at a point in time?

_____

69. How is work performed in a domain?

_____

70. When are activities in domains similar and often repetitive?

_____

71. Regardless of the organization's size, industry or business focus, and/or geographic location, what do the domains reflect?

_____

72. Where do programs originate?

_____

73. What happens during an organization's portfolio review process?

_____

74. When programs are reviewed, what three documents are used to show the current and most suitable profile of the intended outcomes?

_____

75. What happens when the program is approved?

_____

76. What are the two key elements used to clarify the differences between a program and a portfolio?

_____

77. Why is relatedness a consideration on programs?

_____

Answers

68. The nature and complexity of the program.

69. Iteratively and repeated frequently.

70. When organizations pursue similar programs.

71. The higher-level business functions that are essential aspects of the program manager's role.

72. From the organization's strategic planning efforts.

73. Programs are evaluated to see if they still support the organization's strategic goals and objectives and are performing as expected.

74. 1. Business case
    2. Charter
    3. Benefits management plan

75. Funding is approved and allocated, and a program manager is assigned.

76. Relatedness and time.

77. The program work is interdependent to achieve the program's full benefits.

Questions

78. Why is time a key consideration?

_____

79. What four items clarify a program concerning time?

_____

80. Why are schedules important in a program?

_____

81. What is an inevitable challenge of managing a program?

_____

82. When is uncertainty extremely high?

_____

83. Why is there more uncertainty in managing a program versus a project?

_____

84. The program's ability to deal with uncertainty affects the projects in three ways, which are:

_____

85. What is done to ensure the program's outcomes remain in alignment with its intended benefits?

_____

Answers

78. Although a program may span years or even decades, it still is temporary.

---

79. 1. A clearly defined beginning.
    2. A future end point.
    3. A set of outcomes.
    4. Planned benefits to be achieved.

---

80. They enable specific milestone achievements to be realized.

---

81. Uncertainty.

---

82. At the beginning of the program.

---

83. Because of uncertainties in the external environment as well as in the internal environment – with the latter also affecting projects.

---

84. 1. Change the direction of its projects.
    2. Cancel projects.
    3. Start new projects.

---

85. Scope and content are continually elaborated, clarified, and adjusted.

Questions

86. When does additional uncertainty arise?

_____

87. What are two types of change?

_____

88. What is internal change?

_____

89. What is external change?

_____

90. Why are programs better equipped to deal with change than projects?

_____

91. What is needed to manage change on a program?

_____

92. Why is change management a key activity on programs?

_____

93. What happens if there is a change in one component on a program?

_____

94. When is the change management plan developed?

_____

Answers

86. If its components do not contribute to anticipated outcomes even though they meet success criteria and provide outputs, products, and services as planned.

87. Internal change and external change.

88. Changes within the program.

89. The need to adapt the organization to be able to exploit the benefits from the program.

90. Projects are concerned about changes in scope time, and cost; programs can change the direction of a component, cancel one, or start a new component.

91. Strategic insight and understanding of the program's objectives and intended benefits.

92. It enables stakeholders to analyze the need for proposed change, its impact, and the approach or process to implement and communicate change.

93. It may have a direct impact on the delivery of other related components, which then may lead to a need for change on other components.

94. During program preparation.

Questions

95. For best results in managing change, what should the program manager require?

_____

96. Given the program's inherent complexity, what can the program manager do?

_____

97. Using change management to redirect or modify the program roadmap, the program manager should align it with four items. List them.

_____

98. Why do programs use change management in a forward looking, proactive way?

_____

99. What are three sources of complexity in programs?

_____

100. List nine factors that contribute to program complexity.

_____

Answers

95. All components be performed in a way that contribute to the program's outcome and contributed benefits or reduce negative outcomes.

---

96. Group components into other programs to manage them more effectively or redirect, re-plan, or stop components entirely.

---

97. 1. Expected benefits to be delivered.
    2. The new strategy.
    3. The social, regulatory or economic state.
    4. The perception of the program's intended beneficiaries.

---

98. To adapt to an evolving process.

---

99. 1. Human behavior
    2. System behavior
    3. Ambiguity

---

100. 1. Governance
     2. Stakeholders
     3. Definition
     4. Benefits delivery
     5. Interdependencies
     6. Resource
     7. Scope
     8. Change
     9. Risk complexity

Questions

101. What is interdependency complexity?

_____

102. When can these interdependencies occur?

_____

103. What is change complexity?

_____

104. When is change complexity low?

_____

105. When can change complexity be extremely high?

_____

Answers

101. Interdependencies among components need to be defined clearly.

___

102. Outside the program with other projects and programs or external to the organization.

___

103. It arises from the different levels of impact the change from the program can cause the organization.

___

104. When the program changes basic processes in one or two departments.

___

105. When the program transforms a functional organization to one that is project/program oriented.

___

Program Strategy Alignment

Questions

1. What is the definition of Program Strategy Alignment?

_____

2. Since programs align with organizational strategy and to facilitate benefit realization, what do program managers need to know?

_____

3. What precedes an organization's development of strategy?

_____

4. Since programs are to align with organizational objectives and realize benefits, what does the program manager need to do?

_____

5. When is the business case reviewed and what may precede it?

_____

6. What happens after organizational leaders determine the strategy?

_____

7. What happens if the organization is mature in terms of program and project management?

_____

Answers

1. It identifies program outputs and outcomes to provide benefits aligned with the organization's strategic goals and objectives.

---

2. An understanding of how the program will fulfill the portfolio and organization's strategy, goals, and objectives.

---

3. An organization or selection process to determine initiatives to approve, deny, or defer programs.

---

4. Have a thorough understanding of how the program will fulfill the portfolio and organization's strategy.

---

5. During the program formulation sub-phase of the life cycle; a concept may precede it to help develop the business case.

---

6. An initial evaluation and selection process is used to determine the initiatives to approve, deny, or defer programs.

---

7. It will have a formal process for program selection using a Portfolio Review Board (PRB) or Program Steering Committee.

Questions

8. What is the role of the PRB or Program Steering Committee?

_____

9. Who issues the charter?

_____

10. What are four purposes of the charter?

_____

11. What happens after the scope is defined?

_____

12. What are two significant items that happens after resources are confirmed?

_____

13. What is the program manager's role with the individual project management plans?

_____

14. What are five elements of Program Strategic Alignment?

_____

15. What precedes Program Strategic Alignment?

_____

Answers

8. It may issue a program charter defining strategic objectives and benefits the program is selected to deliver.

9. The program sponsor.

10. 1. To authorize the program manager to use organizational resources.
    2. To link the program to the organization's strategic objectives.
    3. To define the program's scope.
    4. To confirm resource commitments.

11. It is presented to governance for approval, funding, and authorization.

12. 1. The program is evaluated to see if it is the best approach to achieve objectives.
    2. The program definition phase begins.

13. To ensure alignment with the program's goals and its intended benefits.

14. 1. Program Business Case
    2. Program Charter
    3. Program Roadmap
    4. Program Risk Management Strategy
    5. Environmental Assessments

15. Organization's strategic plan.

Questions

16. What follows Program Strategic Alignment?

_____

17. What initiates Program Strategic Alignment?

_____

18. What are three key purposes of the program's business case?

_____

19. What is another term for the program's business case?

_____

20. When is the program's strategy alignment process initiated?

_____

21. When is the program strategy alignment process completed?

_____

22. What two items happen during this time?

_____

23. What happens if misalignment is identified?

_____

24. When is misalignment likely to occur?

_____

Answers

16. Program management plan.

17. The program business case.

18. 1. It establishes the validity of the benefits to be delivered.
    2. It defines how outcomes support organizational goals and objectives.
    3. It is an input to the program charter and the roadmap.

19. A documented economic feasibility study.

20. During the program formulation sub-phase.

21. When the program ends.

22. 1. Management processes are set up for economic factors, outcomes, and benefits.
    2. Processes to manage and control risks are set up within the governance framework.

23. The program management plan or the organization's goals and objectives are revised for alignment.

24. In research as results take time, and the organization may change its strategy to better leverage program results.

Questions

25. Why do organization's build strategy?

_____

26. What happens when the strategic planning cycle is updated?

_____

27. List the seven initiatives why the strategic plan is subdivided.

_____

28. What happens with these initiatives?

_____

29. Why is the strategic plan further delineated?

_____

30. List five measurable elements that may be used as the strategic plan is delineated.

_____

31. What is the goal of linking the program to the organization's strategic plan?

_____

Answers

25. To define how their vision will be achieved.

---

26. The organization's strategic goals and objectives are documented in the strategic plan.

---

27. 1. Market dynamics
    2. Customer and partner requests
    3. Shareholders
    4. Government regulations
    5. Organizational strengths and weaknesses
    6. Risk exposure
    7. Competitor plans and actions

---

28. They may later be converted into portfolios.

---

29. To facilitate alignment and goal setting.

---

30. 1. Products
    2. Deliverables
    3. Benefits
    4. Cost
    5. Timing

---

31. To plan and manage the program to help the organization achieve the strategic goals and objectives and to balance the use of resources while maximizing value.

Questions

32. How is the goal of linking the program to the organization's strategic plan achieved?

_____

33. Who does the program manager collaborate with to develop the business case?

_____

34. Why is the business case developed?

_____

35. What does the business case usually describe?

_____

36. What are two ways to present the business case?

_____

37. List the 17 items generally in a business case.

_____

Answers

32. Through the business case.

___

33. Key sponsors and stakeholders.

___

34. To assess the program's investment against intended benefits.

___

35. Key parameters to use to assess objectives and constraints for the intended program.

___

36. As basic and high level or as detailed and comprehensive.

___

37.
   1. Program Outcomes
   2. Approved Concept
   3. Issues
   4. High-level Risk and Opportunity Assessment
   5. Key Assumptions
   6. Business and Operational Impact
   7. Cost/Benefit Analysis
   8. Alternative Solutions
   9. Financial Analysis
   10. Intrinsic and Extrinsic Benefits
   11. Market Demands or Barriers
   12. Potential Profits
   13. Social Needs
   14. Environmental Influences
   15. Legal Implications
   16. Time to Market Constraints
   17. Extent of Alignment to the Strategic Plan

___

Questions

38. List two other purposes of the business case.

_____

39. What is a required document before the program is chartered?

_____

40. What is the primary document for an investment decision?

_____

41. What document describes the program's success criteria?

_____

42. How is program success measured?

_____

43. What is the purpose of the program charter?

_____

44. What happens after the charter is prepared?

_____

45. List 10 key elements of the program charter.

_____

Answers

38. 1. Formal declaration of the value the program is expected to deliver.

    2. Justification for needed resources.

---

39. The business case.

---

40. The business case.

---

41. The business case.

---

42. By the variance between the achieved and planned outcomes.

---

43. To define and authorize the program manager and define the program's scope.

---

44. It is presented to governance for approval, funding, and authorization.

---

45. 1. Program scope

    2. Assumptions

    3. Constraints

    4. High-level risks

    5. High-level benefits

    6. Goals and objectives

    7. Success factors

    8. Timing

    9. Key stakeholders

    10. Other provisions in the business case

Questions

46. What document formally expresses the organization's vision, mission, and benefits to be produced by the program?

47. To support the business case, what document defines program goals and objectives aligned with the organization's strategic plan?

48. Assume you wanted to give the program manager authority over other subsidiary programs, projects, and related activities, you would prepare:

49. What is one key document that will be used to measure program success?

50. Assume you want your charter to help measure program success; therefore, you should include in it three items. List them.

51. When planning the program, list four items the program manager analyzes.

52. List three ways the program is defined.

53. What is the program roadmap?

Answers

46. The program charter.

_____

47. The program charter.

_____

48. The program charter.

_____

49. The program charter.

_____

50. 1. Metrics for success.
    2. A method of measurement.
    3. A clear definition of success.

_____

51. 1. Organization's strategic goals and objectives.
    2. Internal and external influences.
    3. Program drivers.
    4. Benefits stakeholders expect from the program.

_____

52. 1. Expected outcomes.
    2. Required resources.
    3. Strategy to deliver changes to implement the new capabilities.

_____

53. A chronological representation of the program's intended direction.

_____

Questions

54. What does the roadmap show graphically?

_____

55. List three was as to how the program roadmap differs from the schedule.

_____

56. How does the roadmap enable more effective program governance?

_____

57. Should the roadmap show component status?

_____

58. Why is the roadmap useful to stakeholders?

_____

59. What are environmental assessments?

_____

60. What are two examples of influences outside of the organization?

_____

61. Why should program managers be concerned about them?

_____

62. Why are enterprise environmental factors external to the program of interest?

_____

Answers

54. Dependencies between major milestones and decision points.

55. 1. Outlines program events for planning a defined schedule.
    2. Shows the times when benefits are realized.
    3. Serves as the basis for transition and integration of new capabilities.

56. It shows how benefits are delivered in stages or milestones.

57. Yes – including their details, durations, and contributions to benefits.

58. It is an effective way to communicate the overarching plan and benefits to build and maintain advocacy.

59. Internal or external program influences that may influence program success.

60. Ones that are internal to the entire organization or from sources external to it.

61. To ensure ongoing stakeholder alignment, the program's alignment to organizational goals and objectives, and overall program success.

62. They may influence the selection, design, funding, and management of it.

Questions

63. Are the enterprise environmental factors under the program manager's control?

___

64. What is the effect of the enterprise environmental factors on the organization's strategic goals?

___

65. If the program becomes misaligned with the organization's new strategic goals, what is the result?

___

66. List 18 examples of enterprise environmental factors.

___

67. Why should these 18 factors be considered?

___

Answers

63. No, which is why they are a concern.

___

64. They may change, and they may lead to the program becoming misaligned with the new goals.

___

65. It may be changed, put on hold, or canceled regardless of performance.

___

66. 1. Business environment
    2. Market
    3. Funding
    4. Resources
    5. Industry
    6. Health, safety, and environment
    7. Economy
    8. Cultural diversity
    9. Economic diversity
    10. Regulatory
    11. Legislative
    12. Growth
    13. Supply base
    14. Technology
    15. Political influence
    16. Audits
    17. New business processes, standards, and practices
    18. Discoveries and inventions

___

67. They help with the ongoing assessment and evolution of the organization and the program's alignment with its goals.

___

Questions

68. How often should the environmental factors be assessed?

_____

69. How does considering the results from one or more environmental analyses assist the program manager?

_____

70. Are analysis and comparison against real or hypothetical alternatives in a business case?

_____

71. What is comparative advantage analysis?

_____

72. How are feasibility studies used?

_____

73. What is used as the base to conduct a feasibility analysis?

_____

74. How does the feasibility analysis assist the decision makers?

_____

75. Why is a SWOT analysis conducted?

_____

76. What part of a SWOT analysis serves as an input to the program risk management strategy?

_____

Answers

68. Throughout the program.

---

69. It highlights factors that could impact the program and informs risk management.

---

70. Yes, for comparison purposes.

---

71. A type of environmental analysis in which what-if analysis is used to show how the program and its objectives and intended benefits could be achieved by other means.

---

72. To assess the program's feasibility against the organization's financial, sourcing, complexity, and constraint profile.

---

73. The business case.

---

74. By contributing to the body of knowledge to approve, defer, or deny the proposed program.

---

75. It helps optimize the program charter and program management plan.

---

76. Analysis of weaknesses and threats.

Questions

77. List two other documents where a SWOT analysis contributes.

_____

78. What are assumptions?

_____

79. When are assumptions first identified?

_____

80. What aspects of the program are affected by assumptions?

_____

81. When do program managers identify and document assumptions and how often?

_____

82. Are assumptions validated?

_____

83. What is historical information analysis?

_____

84. List seven items that are part of historical analysis

_____

Answers

77. 1. The feasibility study.
    2. The business case.

---

78. Factors which are considered true, real, or certain for planning purposes.

---

79. As the business case is prepared.

---

80. All aspects of the program.

---

81. During the planning process and throughout the program as they are progressively elaborated.

---

82. Yes, throughout the program to see they have not been invalidated by events or other program activities.

---

83. Using previously completed programs or phases of ongoing programs as a source for lessons learned and best practices for the program.

---

84. 1. Artifacts
    2. Metrics
    3. Risks
    4. Estimates
    5. Successes
    6. Failures
    7. Lessons learned

Questions

85. Why is a program risk management strategy important?

_____

86. What are the seven program risk management activities to ensure the program is aligned with organizational strategies?

_____

87. List four items that comprise a risk strategy.

_____

88. What is a risk threshold?

_____

89. List three examples of risk thresholds.

_____

90. Why should the risk threshold be identified early in the program?

_____

91. Who is responsible for ensuring the risk thresholds are established in the program and observed?

_____

Answers

85. It facilitates successful delivery of the roadmap and alignment to organizational strategy considering the environmental factors.

---

86. 1. Actively identifying risks.
    2. Monitoring risks.
    3. Analyzing risks.
    4. Accepting risks.
    5. Mitigating risks.
    6. Avoiding risks.
    7. Retiring program risks.

---

87. 1. Defining risk thresholds.
    2. Performing an initial program risk assessment.
    3. Developing a high-level risk response strategy.
    4. Determining how risks are communicated to higher-level organizational leaders.

---

88. The measure of the degree of accepted variation around a program objective reflecting the risk appetite of the organization and its stakeholders.

---

89. 1. Minimum
    2. Qualitative or quantitative
    3. Maximum

---

90. It links program risk management to strategy alignment.

---

91. Program governance and the program management team.

Questions

92. Why is the initial program risk assessment important?

_____

93. List five other reasons why initial risk assessment is important.

_____

94. What happens after the initial risk assessment is performed?

_____

95. What is a program risk response strategy?

_____

96. How are risk thresholds used to identify the response strategy?

_____

97. What happens once the response strategy is established?

_____

98. How does the risk strategy assist in communication?

_____

99. What is the end result of Program Strategy Alignment?

_____

Answers

92. It provides an opportunity to identify risks to organizational strategy alignment.

---

93. 1. Program objectives may not be supportive of organizational objectives.
    2. Program roadmap may not be aligned with the organizational roadmap.
    3. Program roadmap may not be supportive of the portfolio roadmap.
    4. Program objectives may not be supportive of the portfolio objectives.
    5. Program resource requirements may be out of synch with capacity and capability.

---

94. The next step is the development of a risk response strategy.

---

95. It combines the risk thresholds with the initial risk assessment into a plan to manage risks effectively and consistently.

---

96. It is based on rating criteria to show the risk threshold and whether it is a significant risk or the risk rating to lead to a specific response strategy.

---

97. It drives consistency and effectiveness in risk management activities as part of program integration.

---

98. In terms of managing risks consistency through governance.

---

99. It results in a program plan aligned with organizational goals and objectives.

Program Benefits Management

Questions

1. What is the definition of Program Benefits Management?

_____

2. What are five ways that programs deliver benefits?

_____

3. How are benefits delivered to the sponsoring organization?

_____

4. How do programs deliver benefits?

_____

5. How does managing a program enhance benefits delivery?

_____

6. What is the major reason why programs are conducted?

_____

7. What is the primary difference between components and programs in terms of benefits?

_____

Answers

1. It defines, creates, maximizes, and delivers the program's benefits

---

2. 1. To enhance current capabilities.
   2. To facilitate change.
   3. To create or maintain assets.
   4. To offer new products or services.
   5. To develop new opportunities to generate or preserve value.

---

3. As outcomes providing utility to the organization and the program's intended beneficiaries or stakeholders.

---

4. Through their component projects and subsidiary programs that are performed to produce outputs and outcomes.

---

5. It ensures strategies and work plans of the program's components are responsively adapted to component outcomes or to changes in the sponsoring organization's direction or strategies.

---

6. To deliver benefits to the sponsoring organization or its constituents,

---

7. Within a program, the strategies for delivering benefits may need to be optimized adaptively since component outcomes are individually realized.

Questions

8. What is the iterative pursuit of components expected to produce in terms of benefits?

___

9. Where does benefit consolidation and sustainment fit in the life cycle?

___

10. What is an example of a program with incremental benefits?

___

11. What is an example of a program that delivers benefits all at once?

___

12. How are benefits delivered in non-commercial organizations?

___

13. How are benefits delivered in commercial organizations?

___

14. Since benefit delivery and the program's outputs and outcomes from its components may be uncertain, unpredictable, and uncontrollable, what should the program manager do?

___

15. How do program and project managers work together to support the organization?

___

16. What changes can be made to ensure the optimal delivery of benefits?

___

17. What is the purpose of Program Benefits Management?

___

Answers

8. A stream of outputs and outcomes that contribute to organizational benefits.

9. In Program Delivery as part of outputs and outcomes.

10. An organization-wide process improvement program.

11. A drug development program.

12. In the form of societal value such as improved health, safety, or security.

13. In terms of business value.

14. Manage in a way to adapt strategies and plans during the program to optimize benefits delivery.

15. To enable the delivery of benefits required or desired by the organization.

16. Create new components or cancel existing ones.

17. To focus program stakeholders on the outcomes and benefits to be provided by the activities during the program.

Questions

18. List five ways the program manager uses Program Benefits Management continually.

_____

19. What is a benefit?

_____

20. What are two examples of benefits that are concrete and relatively certain?

_____

21. What are three examples of benefits that are difficult to quantify and may produce uncertain outcomes?

_____

22. What are three examples of benefits that may be realized by the organization?

_____

23. What are three benefits that are due to regulatory issues?

_____

Answers

18. 1. Identify and assess the value of the program's benefits.
    2. Monitor interdependencies among component outcomes to ensure their outputs contribute to the overall program benefits.
    3. Analyze the potential impact on the program's changes on the expected benefits and outcomes.
    4. Align expected benefits with the organization's strategic goals and objectives.
    5. Assign accountability for benefit realization and ensure the benefits can be sustained.

---

19. Gains and assets realized in the organization and by stakeholders based on outcomes delivered by the program.

---

20. Achievement of the organization's financial objectives or creating products or services for consumption or utility.

---

21. An improvement in employee morale, improved customer satisfaction, or the reduced incident of a disease or health condition.

---

22. 1. Expanded marked share
    2. Improved financial performance
    3. Operational efficiencies

---

23. 1. Compliance
    2. Avoiding fines
    3. Avoiding adverse publicity

Questions

24. What are five examples of customers and beneficiaries external to the organization?

_____

25. Can benefits be shared among multiple stakeholders?

_____

26. How can the performing organization benefit from an improved capability?

_____

27. What are two examples of negative impacts from a benefit?

_____

28. Why is it necessary to manage negative impacts?

_____

29. When a program manager addresses a negative consequence, list three departments he or she should consult.

_____

30. How long can benefits be realized?

_____

31. How is the program roadmap used to show benefits?

_____

Answers

24. 1. A group of interested parties.
    2. Business sector.
    3. An industry.
    4. A particular demographic.
    5. General population.

---

25. Yes, since they are defined in terms of intended beneficiaries.

---

26. By showing the ability to deliver consistently and sustaining the products, services, and capabilities produced.

---

27. A reduction in personnel or consolidation of positions.

---

28. They are just as important as realizing the benefits and should be managed, measured, and communicated to leadership and key affected stakeholders.

---

29. 1. Legal
    2. Marketing
    3. Human Resources

---

30. After program closure.

---

31. It is a graphical representation of incremental benefits, so it provides a visual to show when ROI might fund future benefits and outcomes.

Questions

32. What two items should be done when incremental benefits are produced?

_____

33. Provide five examples of programs that have deliver benefits at the end of the program.

_____

34. What happens in Program Benefits Management in the program benefit delivery phase?

_____

35. In the program delivery phase, how are benefits planning and analysis activities performed?

_____

36. Are benefits part of the program's deliverables?

_____

37. Is a risk structure for benefits needed?

_____

38. How does probability relate to program benefits?

_____

39. What are two factors to consider when assigning a risk probability?

_____

Answers

32. Prepare intended benefits for the resulting change and be able to sustain the benefits as long as possible.

---

33. 1. Public works programs.
    2. Major construction efforts.
    3. Aerospace programs.
    4. Aircraft manufacturing or shipbuilding.
    5. Medical devices and pharmaceuticals.

---

34. Components are planned, developed, integrated, and managed to help deliver of the program's benefits.

---

35. Iteratively as corrective action may be needed to deliver the program's benefits.

---

36. Yes, and they should be monitored and managed.

---

37. Yes, and it is based on the organization's risk appetite and the program's strategic value.

---

38. Each benefit should have an assigned risk probability.

---

39. The number of components involved, and the organization's ability to absorb change and sustain it.

Questions

40. How does Program Benefits Management relate to Program Strategy Alignment and Program Stakeholder Engagement?

_____

41. How does Program Benefits Management relate to Program Governance?

_____

42. In the benefits life cycle, what is included in the Program Definition phase?

_____

43. In the benefits life cycle, what is included in the Program Delivery phase?

_____

44. In the benefits life cycle, what is included in the Program Closure phase?

_____

45. What is the purpose of benefits identification?

_____

46. List three items of information that are analyzed to identify and qualify the benefits.

_____

47. List two activities that comprise benefits identification.

_____

48. What document is the formal declaration of the program's benefits?

_____

Answers

40. Program Strategy Alignment and Program Stakeholder Engagement provide inputs or parameters to the program including vision, mission, strategic goals and objectives, and the business case.

41. Program performance data are analyzed by governance to ensure the program will achieve its intended benefits and outcomes.

42. Benefits Identification and Benefits Analysis and Planning.

43. Benefits Delivery.

44. Benefits Transition and Benefits Sustainment.

45. To identify and qualify the benefits stakeholders expect to receive.

46. 1. Organizational and business strategies.
    2. Internal and external influences.
    3. Program drivers.

47. Identifying and qualifying business benefits

48. The business case.

Questions

49. What is the purpose of the benefits register?

_____

50. How is it developed in benefits identification?

_____

51. What is he role of key stakeholders in the benefits register in benefits identification?

_____

52. When are key performance indicators identified?

_____

53. When are the KPI's quantitative and qualitative measures identified and elaborated?

_____

54. List 10 items that may comprise the benefits register in benefits identification.

_____

55. What document shows the mapping of planned benefits to components?

_____

56. What is the purpose of benefits analysis and planning?

_____

Answers

49. To collect and list the program's planned benefits and to measure and communicate their delivery during the program.

---

50. Based on the business case, the strategic plan, and other program objectives.

---

51. To develop performance measures for each benefit.

---

52. In benefits identification.

---

53. During benefits analysis and planning.

---

54. 1. List of planned benefits.
    2. Map of the planned benefits to the components.
    3. Measurement of each benefit.
    4. KPIs and their thresholds.
    5. Risk assessment and probability.
    6. Status or progress indicators.
    7. Target dates and milestones
    8. Person, group, or organization responsible.
    9. Process to measure progress against the benefits plan.
    10. Tracking and communicating progress to stakeholders.

---

55. The roadmap.

---

56. To prepare the benefits management plan and develop the benefits metrics.

Questions

57. List the five activities in benefits analysis and planning.

_____

58. Why is it important to quantify the incremental benefits in a program?

_____

59. How can meaningful metrics help the program manager and stakeholders?

_____

60. What is an example of timely benefits delivery?

_____

61. What are two examples of quantification of intangible benefits?

_____

62. List six examples of quantification of realized benefits.

_____

63. Why is costs an example?

_____

Answers

57. 1. Prepare the benefits management plan.
    2. Develop and prioritize components and interdependencies.
    3. Define KPIs and quantitative measures.
    4. Establish the program's baseline and communicate the metrics to stakeholders.
    5. Update positive and negative risks to benefits with new information.

---

58. To ensure the full realization of benefits can be measured during the program.

---

59. To determine if the benefits exceed their control thresholds and are delivered in a timely way.

---

60. The date when realization should start.

---

61. Improved morale or perception of the organization.

---

62. 1. Hours saved, profits increase, and objectives achieved.
    2. Attaining cultural, political, or legislative improvements.
    3. Increased market share.
    4. Reduced competitor strength.
    5. Attaining incremental production improvements.
    6. Costs.

---

63. They may continue after closeout as operational costs to sustain the benefits, and the program may not provide additional funds to the organization to cover deferred costs of new benefits.

Questions

64. List five examples of risks to the program's benefits.

_____

65. What should the program manager do if there are positive risks to opportunities?

_____

66. What are two examples of possible opportunities?

_____

67. How can the governance function help determine if benefit achievement is occurring?

_____

68. List four types of analysis that may aid governance in this role.

_____

69. How is the benefit management plan used in the benefit delivery phase?

_____

Answers

64. 1. Stakeholder acceptance.
    2. Transition complexity.
    3. Amount of change to absorb.
    4. Realization of unexpected outcomes.
    5. Other situations specific to the industry.

---

65. Optimize their delivery.

---

66. Allocating critical resources to components or using new technology to reduce the effort of resources required to deliver the benefits.

---

67. To see if changes to the components or the program may be needed.

---

68. 1. Linking benefits to program objectives.
    2. Determining financial expenditures.
    3. Evaluating measurement criteria including KPIs.
    4. Conducting measurement and review points.

---

69. To verify benefits are being realized as planned and to provide feedback to stakeholders and governance to facilitate benefit success.

Questions

70. List seven items in the benefit management plan.

_____

71. How does benefit management support the roadmap?

_____

72. Why is the benefits register updated during benefit analysis and planning?

_____

73. Why is the benefits register reviewed with stakeholders?

_____

74. What is the purpose of the program delivery phase?

_____

75. List four items to show the relationship of risk management to program delivery.

_____

Answers

70. 1. Definitions of each benefit, assumptions, and how to achieve it.
    2. Link of component outputs to planned program outcomes.
    3. Metrics and KPIs for benefits.
    4. Roles and responsibilities for benefit management.
    5. Transition of benefits and capabilities to an operational state.
    6. Use of benefits by those responsible for sustaining them.
    7. Process to manage the benefit management effort.

---

71. By establishing the architecture to map how the components will deliver capabilities and outcomes intended to achieve program benefits.

---

72. To show how program benefits are mapped to components based on the roadmap.

---

73. To define and review KPIs and other metrics used to monitor program performance.

---

74. To ensure the program delivers the expected benefits in the benefits management plan.

---

75. 1. Risks affecting benefits may be realized.
    2. Risks may need updating.
    3. Risks may be obsolete.
    4. New risks may be identified.

Questions

76. List five activities in benefits delivery.

_____

77. List five examples of KPIs to evaluate.

_____

78. List four groups who should receive benefits reports.

_____

79. Why should consistent monitoring and reporting on benefits be done?

_____

80. Why is benefits management iterative?

_____

81. List three reasons when corrective action is needed.

_____

Answers

76. 1. Monitor the environment and benefit realization to ensure it is in alignment with organizational strategic objectives.
    2. Initiate, perform, transition, and close components and manage any interdependencies.
    3. Evaluate opportunities and threats and update the benefit register for any risks affecting the benefits or any risks that are obsolete.
    4. Evaluate KPIs to monitor benefit delivery.
    5. Record progress to the benefit register and report to stakeholders.

---

77. 1. Program financials
    2. Quality
    3. Safety
    4. Compliance
    5. Stakeholder satisfaction

---

78. 1. Program management office
    2. Program steering committee
    3. Program sponsors
    4. Other program stakeholders

---

79. To enable stakeholders to evaluate the health of the program and take action to ensure successful benefits delivery.

---

80. Because there is a cyclical relationship between benefits analysis and planning and benefits delivery.

---

81. 1. From information gained when monitoring the environment.
    2. To modify components to maintain alignment with program results and organizational strategic objectives.
    3. Based on evaluation of program risks and KPIs.

Questions

82. List five performance reasons why components may require modification.

_____

83. What may be the results from the component's corrective actions?

_____

84. Why are the initiation and closure of components significant program milestones?

_____

85. What should be done when a component closes?

_____

86. What is necessary for a benefit to have value?

_____

87. Why should the actual benefit when it is realized be compared to the planned benefit?

_____

88. What should you do if the benefits proposition changes or it is delivered too late?

_____

89. What is an example of a change to the benefit's proposition?

_____

90. Once you update and assess the roadmap, what is the next step?

_____

Answers

82. 1. Financials
    2. Compliance
    3. Quality
    4. Safety
    5. Stakeholder satisfaction

---

83. New components may be added, changed, or terminated during benefit delivery.

---

84. The indicate the achievement and delivery of incremental benefits.

---

85. The roadmap should be updated.

---

86. It needs to be realized to a significant degree and in a timely way.

---

87. To help determine if the components and the entire program remain viable.

---

88. Update and assess the roadmap

---

89. The overall life cycle costs exceed the proposed benefits.

---

90. To see if opportunities to optimize the program's pacing may be needed and/or synergies and efficiencies between the components need evaluation.

Questions

91. What happens if there are changes to the program components and changes to the program?

92. Assume there is a benefits review by governance during the program, what two areas are of particular interest?

93. Why is strategic alignment important during a benefits review in an internal program?

94. List three reasons why strategic alignment is important for any type of program in a benefits review.

95. Why is value delivery important in a benefits review?

96. What are two concerns about value delivery in a benefits review?

97. What is the benefits transition phase?

98. How is value delivered when benefits are transitioned?

99. What are the two key activities in benefits transition?

Answers

91. You need to update the benefits management plan and the roadmap.

92. Strategic alignment and value delivery.

93. To measure the effect of the new benefits on the flow of operations and to minimize any negative impacts or disruptions.

94.   1. Ensure the link between the enterprise and program plans.
    2. Define, maintain, and validate the benefits value proposition.
    3. Align program management with operational management.

95. To ensure the program delivers the intended benefits.

96. To ensure that if there was a window of opportunity it was met and to ensure investments still have time value.

97. It ensures benefits are transitioned to operations and then can be sustained.

98. When beneficiaries use the benefits.

99. Verifying the integration, transition, and program closure meet or exceed benefit realization criteria and developing a transition plan.

Questions

100. List six items that should be done in benefits transition.

_____

101. What is the responsibility of the receiving organization or function in benefits transition?

_____

102. What are two examples of when benefits transition can occur?

_____

103. Why should benefits be quantified?

_____

104. What happens if benefits are not realized and the program has closed?

_____

105. List three reasons concerning the emphasis in benefit transition on the individual components.

_____

Answers

100. 1. Define the scope of the transition.
     2. Identify the stakeholders in the receiving units.
     3. Engage the stakeholders in planning the transition.
     4. Measure the program's benefits.
     5. Develop sustainment plans.
     6. Execute the transition.

---

101. To prepare processes and activities to ensure the benefits are received and incorporated.

---

102. Following the close of a component and the close of the overall program.

---

103. To measure their realization over time.

---

104. They need to be monitored.

---

105. 1. Ensure their results or outputs meet acceptance criteria.
     2. Are closed or integrated into other program elements.
     3. Contribute to the collective program benefits.

Questions

106. List 10 transition acceptance activities.

_____

107. List four examples of benefit receivers.

_____

108. List six items to be provided to the transition receiver?

_____

109. Who monitors any remaining risks to the benefits?

_____

110. What is the purpose of the benefits sustainment phase?

_____

Answers

106. 1. Evaluate program and component performance based on acceptance criteria and KPIs.
    2. Review and evaluate acceptance criteria for delivered components or outputs.
    3. Review operational and program process documents.
    4. Review training and maintenance activities.
    5. Review contracts.
    6. Assess if changes are successfully implemented.
    7. Improve acceptance of resulting changes.
    8. Transfer remaining risks to the receiving organization.
    9. Assess readiness and approval by the receivers.
    10. Dispose related resources.

107. 1. For a product line – a product support unit.
    2. For a service – the service management organization.
    3. For an external customer – the customer.
    4. Another program.

108. 1. All key documents
    2. Training and materials
    3. Support systems
    4. Facilities
    5. People
    6. Transition meetings and conferences

109. A governance organization or a PMO.

110. The ongoing maintenance activities after the program ends to ensure continued use of the program's improvements and outcomes.

Questions

111. When should the benefits sustainment plan be prepared?

_____

112. List five items identified in the sustainment plan.

_____

113. Who should be involved in preparing the sustainment plan?

_____

114. Who is responsible for the post-transition activities?

_____

115. Even if the ongoing product, service, or capability support activities are within the scope of the program, how are they managed?

_____

116. List 13 activities in benefit sustainment.

_____

Answers

111. Prior to program closure

---

112. 1. Risks
      2. Processes
      3. Measures
      4. Metrics
      5. Tools

---

113. The program manager and component project managers.

---

114. The program manager.

---

115. They are operational and thus are not run as a program or project.

---

116. 1. Planning for the changes for program recipients to be able to monitor requirements.
      2. Implementing the change efforts.
      3. Monitoring performance for reliability and availability for use.
      4. Monitoring suitability to provide expected benefits to the new owners.
      5. Monitoring availability of logistics support.
      6. Responding to customer inputs or support assistance.
      7. Providing on-demand support as needed.
      8. Planning for and establishing operational support.
      9. Updating technical information.
     10. Planning the transition from program management to operations.
     11. Planning the retirement or phase out of the product or service.
     12. Developing business cases for needed new projects or programs because of operational issues.
     13. Monitoring outstanding risks.

Program Stakeholder Engagement

Questions

1. What is the definition of Program Stakeholder Engagement?

_____

2. What is stakeholder complexity?

_____

3. What is the definition of complexity?

_____

4. What is a stakeholder?

_____

5. Why should program managers be aware of stakeholders' impact and level of influence?

_____

6. Why can stakeholders not be managed directly?

_____

7. Why is it important to balance stakeholder interests?

_____

Answers

1. Identifies and analyzes stakeholder needs and manages expectations and communications to foster stakeholder support.

2. It is due to the differences in stakeholder needs and influence, the number of stakeholders, and it also focuses on the program team and its diversity.

3. Since programs result in change, the definition of complexity focuses on agreement of the future state by stakeholders.

4. An individual, group, or organization that is affected by or perceived to be affected by a decision, activity, or outcome of a project, program, or portfolio.

5. To understand and address the changing environment of the program and its component projects.

6. They are not program resources, so you can only manage their expectations, and people often resist direct management if the relationship is not hierarchical.

7. Because of their potential impact on benefits realization or the inherent conflicting nature of their interests.

Questions

8. How is stakeholder engagement often expressed?

_____

9. What are five other ways in addition to communications to engage stakeholders?

_____

10. What is the primary objective of stakeholder engagement?

_____

11. What are two common characteristics of programs?

_____

12. Why should the program manager map stakeholders?

_____

13. What is the program manager's responsibility concerning stakeholders?

_____

14. List four ways program managers interact with stakeholders.

_____

15. How can the program manager involve stakeholders in the program?

_____

Answers

8. As direct and indirect communication between the stakeholders and the program manager and his or her team.

9. In goal setting, quality analysis reviews, negotiating objectives, agreeing on desired benefits, and committing to resources and ongoing support.

10. To gain and maintain buy in to the program's objectives, benefits, and outcomes.

11. Uncertainty and ambiguity.

12. To ensure successful expectation management and then deliver business benefits to the organization.

13. To spend sufficient time and energy with known stakeholders to ensure all points of view are considered and addressed.

14. 1. Assess attitudes and interests toward the program and their readiness for change.
    2. Involve them in program activities.
    3. Monitor their feedback to the program.
    4. Support training initiatives.

15. Target communications to their needs, interests, requirements, expectations, and wants considering change readiness and the organization's change strategy.

Questions

16. Why does the program manager want a two-way communication with stakeholders?

17. Why is stakeholder engagement at the program level challenging?

18. What are four reasons why people resist change?

19. Why is the program manager the champion for change?

20. Why is it important to understand stakeholders' agendas?

21. Why should the program manager work with the sponsor and governance?

22. How can the program manager bridge the gap between the current state and desired future state?

23. In moving to the organization's desired future state, what is a key competency for each program manager?

24. Why are strong leadership skills needed by the program manager in stakeholder engagement?

Answers

16. To deliver benefits for the organization according to the program charter.

17. Because some stakeholders view the program's benefits as change.

18. They have not requested it, have not participated in creating it, do not understand the need for it, or are concerned about its effect on them personally.

19. The program manager should understand the agendas of stakeholders during the program.

20. Stakeholders could attempt to alter the course of the program or intentionally derail it.

21. To enable effective realization of program benefits.

22. By understanding how the program and its benefits will help the organization move to the future state.

23. Familiarity with organizational change management.

24. To set stakeholder engagement goals to address the changes the program will bring.

Questions

25. List five stakeholder engagement goals.

_____

26. What is program stakeholder identification?

_____

27. List seven items that comprise a stakeholder register.

_____

28. What are three examples of other characteristics in a stakeholder register?

_____

29. How should the stakeholder register be established and maintained?

_____

30. Why should the stakeholder register be maintained in a secure location?

_____

31. When should the program manager comply with data privacy regulations?

_____

Answers

25. 1. Assess readiness for change.

    2. Plan for the change.

    3. Provide resources and support for the change.

    4. Facilitate or negotiate the approach to implement the change.

    5. Obtain and evaluate feedback from stakeholders on the change.

---

26. To systematically identify all key stakeholders and stakeholder groups and list them in a stakeholder register.

---

27. 1. Stakeholder name

    2. Organization position

    3. Program role

    4. Support level

    5. Influence

    6. Communication preference

    7. Other characteristics

---

28. Interests, needs, and status such as engaged.

---

29. In a way the program team can reference it easily for reporting, distributing deliverables, and providing communications.

---

30. It may contain political and legally sensitive communications.

---

31. Based on the country where the program operates.

Questions

32. Why is the stakeholder register a dynamic document?

_____

33. List 15 examples of key program stakeholders.

_____

34. What is the purpose of the program steering committee?

_____

35. What is another name for the program steering committee?

_____

36. Why is the customer a major stakeholder?

_____

37. What are examples of other groups as stakeholders?

_____

Answers

32. New stakeholders may be identified, and the interests of existing stakeholders may change.

---

33. 1. Program sponsor
    2. Program steering committee
    3. Portfolio manager
    4. Program manager
    5. Project manager
    6. Program team members
    7. Project team members
    8. Funding organization
    9. Performing organization
    10. PMO
    11. Customers
    12. Potential customers
    13. Regulatory agencies
    14. Affected individuals or organizations
    15. Other groups

---

34. It is a group of people who represent various program-related interests to provide guidance, endorsements, and approvals through governance practices.

---

35. The program governance board.

---

36. The customer when the program is complete will influence whether it is a success.

---

37. Groups representing consumer, environmental, or other interests, including political influences.

Questions

38. What is a technique used to identify stakeholders across the program life cycle?

_____

39. What is a tool that leads to effective stakeholder engagement?

_____

40. What happens once the stakeholders are listed in the register?

_____

41. What is the purpose of categorizing the stakeholders?

_____

42. After categorizing the stakeholders, what type of information should the program manager obtain from stakeholders?

_____

43. What are four ways to obtain information from stakeholders?

_____

44. How should key information be gathered?

_____

45. What happens after stakeholder information is gathered?

_____

46. Why is the prioritized list needed?

_____

47. How can the program manager effectively use this list?

_____

Answers

38. Brainstorming.

39. The stakeholder register.

40. The program manager categorizes them and begins to analyze them.

41. To highlight differences in needs, expectations, or influence.

42. Views on the organization's culture, politics, concerns, and overall impact of the program.

43. Historical information, individual interviews, focus groups, or questionnaires or surveys.

44. By using open-ended questions.

45. A prioritized list of stakeholders is developed.

46. To focus stakeholder engagement on those with the most influence on the program – both positive and negative.

47. To balance activities to mitigate the effect of stakeholders who are negative and encourage active support from ones who are positive.

Questions

48. When should a stakeholder map be used?

---

49. What is the purpose of the stakeholder map?

---

50. How does the map help the program team make decisions?

---

51. What is another classification model to consider?

---

52. What is represented in the power/interest grid?

---

53. How can the program manager create a framework to address ongoing program activities and evolving stakeholder needs?

---

54. How can potential partnerships among stakeholders be identified?

---

55. How can the program manager determine when to engage stakeholders at various times?

---

56. When should the stakeholder register and prioritization analysis be reviewed?

---

57. What is stakeholder engagement planning?

Answers

48. On complex programs.

___

49. To visually present the interaction of the stakeholders' current and desired support and influence.

___

50. It shows how and when to engage stakeholders considering their interest, influence, involvement, interdependencies, and support.

___

51. A power/interest grid.

___

52. The stakeholders' level of authority or power and their level of concern or interest.

___

53. By identifying stakeholder expectations and outlining KPIs and expected benefits.

___

54. Through the stakeholder map as it can show collaboration opportunities as well.

___

55. Through the stakeholder map, he or she can remind teams of which stakeholders need to be engaged throughout the program.

___

56. Regularly, as different stakeholders will be identified, and others may have different levels of interest or influence at different times in the program.

___

57. It outlines how all stakeholders will be engaged during the program.

___

Questions

58. To understand the program's environment to best engage stakeholders, list five key documents to analyze.

_____

59. List six areas to consider during stakeholder engagement and planning.

_____

60. What is the result of stakeholder engagement and planning?

_____

61. What is the purpose of the stakeholder engagement plan?

_____

62. What are two areas covered in a stakeholder engagement plan?

_____

63. What are three examples of metrics in a stakeholder engagement plan?

_____

64. Are the stakeholder engagement guidelines provided to program components?

_____

Answers

58. 1. Stakeholder register
    2. Stakeholder map
    3. Organization's strategic plan
    4. Program charter
    5. Business case

59. 1. Organization's culture and acceptance of change.
    2. Attitude about the program and sponsors.
    3. Phases applicable to stakeholder engagement.
    4. Expectations of benefit delivery.
    5. Support or opposition to the benefits.
    6. Ability to influence the program.

60. The stakeholder engagement plan.

61. To have a detailed strategy for effective stakeholder engagement based on the situation.

62. Guidelines and insight as to how stakeholders are engaged, and metrics used to measure performance of stakeholder engagement activities.

63. 1. Participation in meetings and other communications channels.
    2. Degree of active or passive support or resistance.
    3. Effectiveness of engagement to meet its intended goal.

64. Yes, they are given to projects, subsidiary programs, or other program activities.

Questions

65. How does the stakeholder engagement plan support other parts of the program?

_____

66. What is program stakeholder engagement?

_____

67. What is one of the primary roles of the program manager?

_____

68. How can the program team communicate program benefits and how they relate to the program's strategic objectives?

_____

69. What are three skills the program manager may use to defuse stakeholder opposition to the program?

_____

70. When are facilitated negotiation sessions needed?

_____

71. What types of information should the program manager provide to help stakeholders have common expectations about the program?

_____

72. What are three primary metrics for stakeholder engagement?

_____

73. What are four items that should be logged concerning stakeholder engagement?

_____

Answers

65. It provides information used to develop program documentation and ongoing alignment since stakeholder change throughout the program.

66. A continuous activity since stakeholders change during the program and their attitudes change when the program progresses and delivers benefits.

67. To ensure all stakeholders are adequately and appropriately engaged.

68. By interacting and engaging with stakeholders.

69. Communication, motivation, and conflict resolution.

70. On large programs with diverse stakeholders when their expectations conflict.

71. The charter and business case and possibly an executive summary to summarize risks, dependencies, and benefits.

72. Positive contributions to benefits realization, stakeholder participation, and the communication frequency or rate with the program team.

73. Meeting invitations, attendance, meeting minutes, and action items.

Questions

74. Why should program managers review stakeholder metrics regularly?

_____

75. Why should participation trends be analyzed?

_____

76. Why should the history of stakeholder participation be analyzed?

_____

77. Why is thorough stakeholder engagement analysis useful?

_____

78. How can use of an issue log help the program team in stakeholder engagement?

_____

79. What should the program manager use to track issues on a small program?

_____

80. List seven items that may be affected because of stakeholder issues and concerns.

_____

81. When should impact analysis be used in stakeholder engagement?

_____

Answers

74. To identify risks if stakeholders are not participating.

75. By using root-cause analysis, the program manager can identify the causes of not participating.

76. It can provide background information to influence stakeholder perceptions and expectations.

77. It can show any incorrect assumptions that could lead to unanticipated issues or poor program management decisions.

78. By documenting, prioritizing, and tracking issues, the program team can better understand stakeholder feedback.

79. A simple spreadsheet.

80. 1. Scope
    2. Benefits
    3. Risks
    4. Cost
    5. Schedule
    6. Priorities
    7. Outcomes

81. To recognize the urgency and probability of stakeholder issues and determine if any are risks to the program.

Questions

82. What is program stakeholder communications?

_____

83. What is considered to be the 'heart' of stakeholder engagement?

_____

84. What is the key to executing the program and delivering its benefits?

_____

85. Why should a strategy be devised for each stakeholder listed in the register regarding communications?

_____

86. Why should the program manager use a communications feedback loop?

_____

87. Why is a defined communications process important in stakeholder engagement?

_____

88. What should be done if stakeholders raise questions?

_____

89. What information should be available to decision makers?

_____

90. Why should the program manager continually monitor the environment?

_____

Answers

82. It creates a bridge between diverse stakeholders with different cultural and organizational backgrounds, levels of expertise, perspectives, and interests.

83. Communications

84. Communications

85. To determine communication requirements such as language, format, content, and the amount of detail needed.

86. To discuss program changes and escalation processes.

87. To target stakeholder support for the program's approach and the delivery of benefits.

88. Capture them and their answers in a log so others can benefit.

89. Information formatted to meet their needs so decisions can be made at the right time to move the program forward.

90. To ensure stakeholder communications needs are met.

Program Governance

Questions

1. What is the definition of Program Governance?

2. What is governance complexity?

3. What does program governance comprise?

4. What is the focus of program governance?

5. What types of practices does the program governance framework provide?

6. How is program governance performed?

7. What is the program manager's responsibilities concerning the governance framework?

Answers

1. Enables and performs program decision making, establishes practices to support the program, and maintains program oversight.

---

2. It results from the program sponsor and the related components' sponsors, management structures, and the program's decision-making processes.

---

3. The framework, functions, and processes to monitor, manage, and support the program so it meets strategic and operational goals.

---

4. To deliver program benefits by establishing systems and methods such that the program and its strategy are defined, authorized, monitored, and supported by the organization.

---

5. It assists with decision making and ensures the program is managed appropriately.

---

6. By a review and decision-making group charged with endorsing or approving recommendations for a program.

---

7. To ensure the program follows the framework and to manage its day-to-day activities.

Questions

8. What is the role of the program team concerning governance?

_____

9. How is governance of the components often achieved?

_____

10. If the program manager has governance of the components, what is it called?

_____

11. What is organizational governance?

_____

12. What is the hierarchical level where program investments are authorized?

_____

13. How is portfolio governance linked to programs?

_____

14. What happens in terms of governance with a standalone program outside of a portfolio structure?

_____

15. Who determines the type and frequency of governance activities?

_____

16. If governance is handled by the portfolio, what does it provide to the program?

_____

Answers

8. The program manger ensures the team understands and follows the governance procedures and underlying governance principles.

9. By the program manager and program team responsible for the program's integrated outcomes.

10. Component governance

11. A structured way to provide control, direction, and coordination through people, processes, and policies to meet strategic objectives and operational goals

12. Portfolio governance

13. Through governance

14. A governing board provides governance-supporting functions and procedures to the program.

15. Portfolio governance and the governance board.

16. Governance policies, oversight, control, integration, and decision-making functions and processes.

Questions

17. List ten ways governance supports program success.

_____

18. What else is involved in terms of ensuring the program is in alignment with the organization's goals?

_____

19. What else is involved in establishing agreements?

_____

20. How is governance involved in facilitating stakeholder engagement?

_____

21. What happens to ensure the program is compliant with portfolio and corporate governance policies and processes?

_____

22. Why is effective governance required if the program is operating in a complex or uncertain environment?

_____

23. Why are governance decisions important?

_____

Answers

17. 1. Alignment of goals with the organization's strategic goals.
    2. Approve, endorse, and initiate the program, including funding.
    3. Establish agreements as to how the sponsoring organization will oversee the program.
    4. Facilitate engagement of program stakeholders.
    5. Communicate risks and opportunities to the supporting organization.
    6. Align the program with portfolio and corporate governance policies and processes.
    7. Conduct phase-gate reviews, decision-point reviews, and program health checks.
    8. Assess the validity of the organization's strategic plan and level of support.
    9. Endorse pursuit of components.
    10. Make decisions between phases and to terminate or close the program.

---

18. Alignment with the strategic vision, organizational capabilities, and resource commitments of the sponsoring organization as well as compliance with reporting and controlling processes.

---

19. Determining the degree of autonomy that the program will be given to pursue its goals.

---

20. By establishing clear expectations for the program's interaction with key governing stakeholders.

---

21. A program may need to create a particular governance process and procedure that is aligned with the organization's governance principles.

---

22. When it is necessary to respond quickly to outcomes and have information available during the program.

---

23. To focus on facilitating adaptive alignment of the program's approach to enable benefit delivery.

Questions

24. How does governance focus on emergent outcomes?

_____

25. What should you do it the organization has not set up a portfolio in a formal way?

_____

26. What is summarized in a program governance plan?

_____

27. Should the governance plan be a standalone document?

_____

28. Can one governance plan be used for different programs?

_____

29. What is the purpose of a program governance plan?

_____

30. How is the governance plan used during the program?

_____

31. When is the governance plan modified?

_____

32. What is a good practice to follow if the governance plan is modified?

_____

Answers

24. It provides a means for the program to seek authorization to change strategy or plans.

25. Ensure process to develop the idea and steps to authorize the program then are handled by organizational governance.

26. To facilitate the design and implementation of effective governance.

27. It depends – based on the organization, it may be a standalone plan or part of the program management plan.

28. Yes, often organizations have a standard governance plan that is applied to multiple programs.

29. To describe the systems and methods to use to monitor, manage, and support a program; and the responsibilities and roles so there is effective and timely use of the systems and methods.

30. To ensure the program conforms to established governance expectations and agreements.

31. Based on outcomes attained during the program.

32. To communicate the modifications to stakeholders responsible for program governance.

Questions

33. What is in the governance plan concerning roles and responsibilities?

_____

34. What are four examples of planned governance meetings?

_____

35. Should the governance plan contain a schedule of these meetings?

_____

36. What other plan is influenced by the governance plan?

_____

37. List five other items in the governance plan.

_____

38. What are three examples of constraints in the governance plan?

_____

39. Why should metrics and measurements be in the governance plan?

_____

40. How should information on components be collected and reported?

_____

41. What type of information on support services is needed in the governance plan?

_____

Answers

33. Who will have accountability and authority regarding key decision-making capabilities and boundaries.

34. Decision point reviews, phase-gate reviews, program health checks, and required audits.

35. Yes, criteria should be defined for them such as to review the outcomes to influence the program approach or program resource needs.

36. The program management plan as it defines the requirements for governance interaction and review.

37. 1. Dependencies, assumptions, and constraints.
    2. Benefits, performance metrics, and measurements.
    3. Support services.
    4. Stakeholder engagement.
    5. Governance practices.

38. Resources, budget, and operational limitations.

39. To evaluate the program and component contributions to metrics.

40. Through a balanced scoreboard or a dashboard.

41. A description of the feedback and support used during the program.

Questions

42. What information on stakeholder engagement should be in the governance plan?

_____

43. What type of communication practices should be included in the governance plan?

_____

44. Why is a vision for program governance needed?

_____

45. What happens in most organizations concerning governance and program approval, endorsement, and definition?

_____

46. When does governance make these approvals?

_____

47. What two artifacts does the governance board use in making these approvals?

_____

48. Why does the governance board use the business case?

_____

49. Why does the governance board use the program charter?

_____

50. What is the role of governance in terms of funding?

_____

Answers

42. A list of the stakeholders who need engagement and communication about governance activities.

43. The intended design of communications practices.

44. It is used to ensure the program's vision and goals are defined to support those of the organization.

45. Governance outlines responsibility for the program's approach and for achieving goals and authorizes the use of resources.

46. During program formulation.

47. Business case and program charter.

48. It defines the program's proposed benefits and provides justification for required resources to deliver them.

49. It authorizes the program management team to acquire the resources needed and links the program to the business case and the organization's strategic priorities.

50. It facilitates funding to that requested in the approved business case.

Questions

51. What happens if funding is controlled by a separate budgeting process?

_____

52. How are organizational priorities determined?

_____

53. What does program governance do if funding is to be provided by external sources?

_____

54. What are examples of constraints if funds are provided by external sources?

_____

55. What are the program success factors governance establishes?

_____

56. What type of criteria does governance establish?

_____

57. List 11 examples of monitoring and controlling documents.

_____

Answers

51. Funding is provided in a way that is consistent with program needs and organizational priorities.

---

52. Through the portfolio management processes.

---

53. It enters into agreements to secure it.

---

54. Laws, regulations, and other initiatives.

---

55. The minimum successful criteria for a program and how they will be measured, communicated, and endorsed.

---

56. Ones that describe the definition of success, are consistent with stakeholder expectations, and reinforce the program's alignment to deliver sustainable benefits.

---

57. 1. Operational status and program progress.
    2. Required and incurred resource requirements.
    3. Known risks, their response plans, and escalation criteria.
    4. Strategic and operational assumptions.
    5. Benefits realized and expected sustainment.
    6. Decision criteria, tracking, and communications.
    7. Program change control.
    8. Compliance with corporate and legal policies.
    9. Program information management.
    10. Issues and their response plans.
    11. Program funding and financial performance.

Questions

58. What is governance's involvement to ensure there are effective risk and governance practices?

_____

59. What are the two levels where this process operates?

_____

60. How should the requirements for engaging governing stakeholders for effective risk and issue management be handled?

_____

61. How are program risk thresholds for adherence handled?

_____

62. Who establishes these risk thresholds for the program?

_____

63. Who is responsible for program quality?

_____

64. What are four examples of quality measures governance participants may define?

_____

65. Can quality control activities differ at the component level?

_____

Answers

58. To ensure key risks and issues are escalated appropriately and resolved in a timely way.

---

59. 1. Within the program, between component teams, the program management team, and the governance board or program steering committee.
    2. Outside the program between the program management team, the program steering committee, and other stakeholders.

---

60. They should be documented and communicated.

---

61. Based on the organization's risk appetite, working with organizational governance, and the program management team.

---

62. Program governance.

---

63. Governance participants are responsible for defining quality measures and are responsible for reviewing and approving the quality management approach.

---

64. 1. Minimal quality criteria and standards to be applied to program components.
    2. Minimal requirements for component quality planning, quality control, and quality assurance.
    3. Any required program-related quality assurance and control activities.
    4. Roles and responsibilities for program-level quality assurance and quality control activities.

---

65. Yes, as they are based on the component's uncertainty and complexity level.

Questions

66. What is governance's role in program changes?

_____

67. How are governance participants well positioned to assess proposed changes?

_____

68. List three areas regarding changes the program manager should assess.

_____

69. What is the extent to which the program steering committee can authorize a change?

_____

70. What should the program management team do if there is a proposed change?

_____

71. What is the purpose of a decision-point review?

_____

72. How are changes handled at decision point reviews?

_____

73. When do key decision points occur?

_____

74. What are phase-gate reviews?

_____

Answers

66. The program steering committee defines the types of changes the program manager can make on his or her own and ones that require further discussion.

67. Through the monitoring, controlling, and reporting processes.

68. 1. If risks are acceptable or desirable.
    2. If the change is operationally feasible and supported by the organization.
    3. If the changes are significant and requires the program steering committee's involvement.

69. It is bounded by the business case and organizational strategy.

70. Record it, its rationale, and the outcome.

71. Governance conducts these reviews when there is the initiation or completion of a significant part of the program or to approve or disapprove the passing of one part of a program segment to another.

72. Governance facilitates the review or approval of any required changes.

73. At the end of program phases.

74. A review to decide whether to continue to the next phase, to continue with modifications, or to end the program or a component.

Questions

75. List 11 assessments that may be done at decision-point reviews.

_____

76. What are two other examples of program reviews?

_____

77. During the reviews, what is the program steering committee authorized to do?

_____

78. What are three examples of when a program should be terminated because of a review?

_____

79. What determines the frequency of a program review and its requirements?

_____

80. Where should the organization's expectations for governance reviews be documented?

_____

Answers

75. 1. Strategic alignment of the program and its components.
    2. Component outcomes to assess actual versus planned benefits.
    3. Risks to the program.
    4. Resource needs and organizational commitments to fulfill them.
    5. Stakeholder satisfaction with performance.
    6. Compliance with quality or process standards.
    7. Impact of external or environmental developments.
    8. Information critical to strategic prioritization or operational investments.
    9. Issues that require resolution.
    10. Possible program changes to improve performance.
    11. Fulfillment of criteria to exit a phase and move to the next phase.

76. Ones for portfolio management or the budget process.

77. Confirm support to continue the program or make recommendations for adaptive changes to improve the program's likelihood to deliver its planned benefits.

78. 1. It will not be able to deliver its planned benefits.
    2. It cannot be supported at the required investment level.
    3. It cannot be supported based on a portfolio review.

79. It is based on the authority and autonomy of the program team to oversee and manage the program.

80. In the program's governance plan.

Questions

81. How is a program periodic health check used?

_____

82. Why are periodic health checks conducted?

_____

83. Where are their requirements documented?

_____

84. Who approves initiation of program components?

_____

85. Who frequently acts as the proposer for a new component?

_____

86. What two items may be required to initiate a new component?

_____

87. List six items needed to approve a new component.

_____

88. What is the approach used in managing activities in the component?

_____

Answers

81. They are held between decision-point reviews to assess performance and progress in realizing and sustaining benefits.

82. Because the time between decision-point reviews may be too great to not assess program performance.

83. In the program governance plan.

84. The program steering committee.

85. The program manager.

86. Introducing other governance structures to manage and monitor the component and the commitment of organizational resources to complete it.

87. 1. Develop, modify, or reconfirm the business case.
    2. Ensure resources are available.
    3. Define or reconfirm individual accountabilities for the component.
    4. Communicate key information about the new component to stakeholders.
    5. Establish a component-specific quality plan if needed.
    6. Authorize the governance structure to track the component's progress.

88. It depends and may be managed according to practices in the PMBOK®, or it may be managed as in program management.

Questions

89. What happens when a new component is initiated?

_____

90. Is approval needed to transition or close a component?

_____

91. List five items needed in a recommendation to transition or close a component.

_____

92. Who makes the decision to close the program?

_____

93. Who approve the final program report?

_____

94. What two areas are assessed before program closure?

_____

95. Why are some programs terminated?

_____

96. What should be done if a program is terminated?

_____

Answers

89. All program-level documentation and records are updated.

90. Yes, and a review recommendation is needed.

91. 1. Confirm the business case is satisfied or that further pursuit of the component's goals should be discontinued.
    2. Communicate the component closure to key stakeholders.
    3. Ensure component compliance with any required quality plans.
    4. Assess organizational or program-level lessons learned.
    5. Confirm accepted practices for project or program transition or closure are satisfied.

92. The program steering committee.

93. The program steering committee.

94. Whether conditions warranting closure are satisfied and if the closure recommendations are consistent with the current organizational strategy, vision, and mission.

95. Because changes in the organization's strategy or environment mean there are diminished benefits or needs.

96. Follow the closure procedures.

Questions

97. Why is it important to transition program governance to operational governance when a program closes?

_____

98. What is needed in governance that is critical to program management success?

_____

99. Since there are often issues when the needs of one program conflict with that of another program, what should the program manager do?

_____

100. What are three other terms for a program steering committee?

_____

101. Within an organization, how is the relationship between program governance and the program management functions best handled?

_____

102. List six key roles in program governance.

_____

103. Why is the program manager a key role in governance?

_____

104. Why is the project manager a key role in governance?

_____

Answers

97. If it is not done, it will impact benefits realization.

98. Establishing a collaborative relationship between the governance board members.

99. Rely on the program steering committee to resolve issues quickly.

100. Program governance board, oversight committee, or board of directors.

101. By assigning key roles to people in these functions who are viewed as important stakeholders.

102. 1. Program sponsor.
     2. Program steering committee.
     3. Program management office.
     4. Program manager.
     5. Project manager.
     6. Other stakeholders – the portfolio manager of components and operational managers who receive the program's capabilities.

103. The program manager interfaces with the governance function and sponsor and manages the program to ensure delivery of its intended benefits.

104. The project manager interfaces with the program manager and sponsor and manages the delivery of the project's product, service, or result.

Questions

105. When can the program sponsor be on the program steering committee?

_____

106. List five key attributes of a sponsor.

_____

107. What are three typical responsibilities of the program sponsor?

_____

108. If the sponsor is the chair of the program's steering committee, what should he or she expect from the organization?

_____

109. How does the sponsor interact with operations?

_____

110. How are program steering committees usually staffed?

_____

111. How are program steering committee members usually selected?

_____

Answers

105. If the sponsor has a senior role in the organization and its investment decisions and who is vested in the success of the organization's programs.

106. 1. Influence stakeholders
     2. Work across different stakeholder groups
     3. Leadership
     4. Decision-making authority
     5. Communication skills

107. 1. Secure program funding and ensure the program is aligned with the strategic vision.
     2. Enable benefit delivery.
     3. Remove any barriers and obstacles to program success.

108. Time and resources to do the job even if relief is needed from other duties.

109. To drive change so operations can accommodate capabilities delivered by the program and secure positive benefits.

110. By people who are individually or collectively recognized as ones with organizational insight and decision-making authority.

111. For strategic insight, technical knowledge, functional responsibilities, operational accountabilities, operational accountabilities, portfolio management responsibilities, and abilities to represent different stakeholder groups.

Questions

112. How can program steering committee members if appropriately selected be able to address issues or questions that may arise about the program?

_____

113. List 11 responsibilities of a program steering committee.

_____

114. What is involved when the program steering committee provides governance support?

_____

115. What is involved when the program steering committee provides governance resources?

_____

116. What happens in a planning session of the program steering committee?

_____

117. What are the program steering committee members leadership responsibilities?

_____

118. What is the preferred way to define key messages to stakeholders?

_____

119. How is a program oversight committee handled in a small organization?

_____

Answers

112. If they are organizational executives and leaders responsible for supporting the program.

113. 1. Provide governance support for the program.
     2. Provide governance resources.
     3. Ensure program goals align with organizational strategic and operational goals.
     4. Conduct planning sessions.
     5. Endorse or approve recommendations and changes.
     6. Provide oversight and monitoring.
     7. Provide leadership.
     8. Define key messages to communicate to stakeholders.
     9. Resolve expected benefits and benefit delivery.
     10. Approve program closure or termination.

114. It includes oversight, control, and decision-making functions.

115. To oversee and monitor program uncertainty and complexity to achieve benefits delivery.

116. They confirm, prioritize, and fund the program.

117. To make, enforce, carry out, and communicate decisions.

118. To make sure they are consistent and transparent.

119. By a senior-level executive.

Questions

120. What is the best way to provide effective and adaptive governance?

_____

121. What are three examples when programs need to report to multiple oversight committees?

_____

122. What is an example of a program managed in an exceedingly complex organization?

_____

123. What is the PMO's role in governance?

_____

124. What are four responsibilities of the PMO?

_____

125. When are multiple PMOs used?

_____

126. When are PMOs considered as centers of excellence?

_____

127. What are two key positions for a PMO?

_____

Answers

120. By a single committee.

---

121. 1. If they are sponsored jointly by public and private organizations.
    2. If there are collaborations between private organizations that also may compete with each other.
    3. If they are programs in exceedingly complex organizations.

---

122. One in which subject matter experts cannot all be part of a single program steering committee.

---

123. To facilitate governance practices.

---

124. 1. Standardizes governance practices.
    2. Facilitates resource sharing, methodologies, tools, and techniques.
    3. Provides staff trained in governance practices.
    4. May monitor compliance to program management practices.

---

125. On large, complicated, and complex programs.

---

126. When the organization wants a high level of consistency and professionalism in managing and governing programs.

---

127. Change and benefit management specialists.

Questions

128. When can the PMO functions be dedicated to one person?

_____

129. Who grants the program manager authority to make decisions?

_____

130. What are three factors that are considered when granting the program manager decision-making authority?

_____

131. List ten factors that are the program manager's governance-related responsibilities.

_____

132. What is involved in assessing the governance function?

_____

133. When should the program manager escalate risks and issues?

_____

134. How are program goals and benefits delivered?

_____

Answers

128. When the person has an exceptional record in program management and governance management practices or to the program manager.

---

129. The program steering committee.

---

130. 1. Experience.
    2. Size and complexity of the program and its components.
    3. Degree of coordination in managing the program in the context of the larger organization.

---

131. 1. Assess the governance framework.
    2. Oversee program conformance to governance policies and processes.
    3. Manage program interaction with the steering committee and sponsor.
    4. Manage component interdependencies.
    5. Monitor and manage program risks, performance, and communications.
    6. Manage program issues and risks and escalate them if necessary.
    7. Monitor and report on program funding and health.
    8. Assess program outcomes.
    9. Create, monitor, and communicate the program roadmap and key dependencies.
    10. Manage, monitor, and track benefits realization.

---

132. It includes organizational structure, policies, and procedures and in some cases establishes the governance framework.

---

133. If they are beyond the program managers control.

---

134. Through the individual components.

Questions

135. What is component governance?

_____

136. What is the program manager's role in component governance?

_____

137. What is the project manager's role in governance?

_____

138. List six governance responsibilities of a project manager.

_____

139. What is the role of the portfolio manager in program governance?

_____

140. What is the role of functional representatives and product owners in program governance?

_____

141. What is the role of the operations manager in program governance?

_____

142. What can happen as benefits are integrated into the organization?

_____

Answers

135. It is similar to program governance, but the program manager and his or her team are the steering committee members.

---

136. He or she designs the governance framework, functions, and processes.

---

137. The project manager is subject to component governance oversight.

---

138. 1. Manage interactions with the program manage, steering committee, and sponsor.
     2. Oversee project performance.
     3. Monitor and manage performance and communications.
     4. Manage project risks and issues, escalating if needed.
     5. Manage internal and external dependencies.
     6. Engage key stakeholders.

---

139. To ensure the program is selected, prioritized, and staffed following the organization's plan to realize desired benefits.

---

140. To ensure the program's direction is aligned to customers and their potential evolving requirements.

---

141. To receive and integrate the program component capabilities for achieving desired benefits.

---

142. Disruption can happen as a steady state is hard to attain since it is different from the previous environment.

Questions

143. What can the operations manager do to accelerate the change?

_____

144. Who is suitable to best manage the new capabilities and benefits, so they are integrated successfully?

_____

145. If a person who is a business change manager is used, who supports this individual?

_____

146. What are three other roles that may be involved in program governance?

_____

147. How does program governance begin?

_____

148. Why will governance practices differ?

_____

149. What are five different areas that affect governance needs?

_____

150. Why is a sponsor organization useful in implementing governance practices?

_____

151. What are two key reasons why effective governance is useful?

_____

Answers

143. Assign at least one person to manage the change.

144. The sponsor, people from the receiving business areas, the program manager, a project manager, and a specialist in managing business change.

145. A team from the business area impacted.

146. Risk specialist, buyers, and contracting experts.

147. By identifying governance participants and establishing governance practices.

148. They depend on the sector or industry.

149. 1. Political
     2. Regulatory
     3. Legal
     4. Technical
     5. Amount of competition

150. To monitor the program's support of the organization's strategy.

151. To ensure strategic alignment is optimized and the program's proposed benefits are delivered as expected.

Questions

152. What do governance practices provide?

_____

153. Why can the governance design have an impact on the success of a program?

_____

154. List ten factors to consider when optimizing or tailoring program governance.

_____

155. Why is the legislative environment important to consider?

_____

156. Why is the decision-making environment important?

_____

157. What is optimized governance?

_____

Answers

152. A foundation so decisions are made with appropriate justification, and responsibilities and accountabilities are defined and applied.

---

153. Inappropriate governance leads to a false impression of the program's progress, its strategic alignment, and success.

---

154. 1. Legislative environment
    2. Decision-making hierarchy
    3. Optimized governance
    4. Alignment with portfolio and organizational governance
    5. Program delivery
    6. Contracting
    7. Risk of failure
    8. Strategic importance
    9. PMO
    10. Program funding structure

---

155. For programs with influence from changing legislation, which may require direct interaction with legislative authorities.

---

156. It affects where the competencies, authorities, and accountabilities reside as more controlling factors are needed in organizations with limited accountabilities for one's actions, and in others, greater autonomy is given to environments where the program manager and team have more autonomy.

---

157. It means being effective and streamlined to perform governance practices.

Questions

158. How should you determine the degree to which program governance is aligned with organizational governance?

159. When is the need for organizational governance the greatest?

160. Is program governance required more when the benefits are delivered all at once or incrementally?

161. What type of governance is needed when contracting is used?

162. What happens in terms of governance if the program is one with a great risk of failure?

163. What is required in terms of governance when the program is one critical to organizational success?

164. What are PMO roles in terms of governance?

165. What happens in terms of governance when funds are received from outside sources?

166. How does the life cycle phase influence program governance?

167. What should be done once program governance is designed and implemented?

Answers

158. By the number, type, and importance of the program governance's interaction with corporate governance and groups.

159. In the program definition stage.

160. Incrementally as there is regular delivery of benefits meaning constant change in the organization.

161. It is based on the legal agreement in the contract.

162. The governance team will monitor progress and success more diligently.

163. More senior level people will be on the governance team.

164. A centralized PMO may perform governance for all programs in the organization, or a PMO may be established for a specific program.

165. There are implications for the governance design and skills required.

166. The relative importance of different governance practices differs as the program progresses.

167. It should be assessed to ensure that it is adding value and whether any changes are needed.

Program Life Cycle Management

Questions

1. What is the definition of Program Life Cycle Management?

_____

2. What is resource complexity?

_____

3. What are three items that add to resource complexity?

_____

4. What is scope complexity?

_____

5. What else contributes to scope complexity?

_____

6. What are two reasons for risk complexity?

_____

7. What does the program life cycle demonstrate?

_____

Answers

1. Manages program activities required to facilitate effective program definition, program delivery, and program closure

---

2. It involves resource availability at the required level of capability and capacity.

---

3. 1. Adequate funding
   2. Suitable supplies
   3. Materials

---

4. It arises because of the difficulty of clearly defining the program's and its components' deliverables and benefits.

---

5. Managing benefits delivery beyond the lifespan of the components.

---

6. 1. The high level of uncertainty with the extended life cycle.
   2. The uncertainty of the components outcomes and interdependencies.

---

7. The non-sequential nature of program's delivery phase.

Questions

8. What are the three phases of the program life cycle?

_____

9. What are two ways as to how scope is handled on programs?

_____

10. What is the program manager's role in risk and issue management compared to that of the project manager?

_____

11. What is the project manager's role in risk and issue management compared to that of the program manager?

_____

12. What happens in terms of information on a program and its projects in the program delivery phase?

_____

13. What happens in terms of information on a program and its projects in the program definition phase?

_____

14. What happens in terms of information on a program and its projects in the program closure phase?

_____

15. What happens during the program delivery phase?

_____

Answers

8. Program Definition, Program Delivery, and Program Closure

---

9. 1. It encompasses the scopes of its components.
   2. It produces benefits by ensuring the components' outputs and outcomes are delivered in a coordinated and complementary manner.

---

10. To monitor and address issues and risks that may impact program performance and to recognize and embrace new opportunities.

---

11. To focus on managing project issues and risks and identify any issues, risks, and dependencies that may impact other components.

---

12. Information about program's benefits, goals, and strategy flows to its projects; any information on timing, strategies, needs, and constraints flows back from the projects to the program.

---

13. Information about the progress, issues, risks, dependencies, outputs and outcomes flows from the projects to the program; the program communicates regularly to ensure component activities are coordinated and aligned to deliver the program's benefits.

---

14. As projects close, information about the project's outcomes and outputs flows from the project to the program to ensure program benefits are fully realized and sustained.

---

15. 1. Components are initiated, planned, executed, delivered, and closed.
    2. Benefits are delivered, transitioned, and sustained.

Questions

16. What is the program life cycle domain?

_____

17. Why should the program manager establish a consistent set of processes and apply them across phases?

_____

18. When do the program life cycle activities begin?

_____

19. List five items that occur in the program definition phase.

_____

20. List three items in the program delivery phase.

_____

21. List two activities in the program closure phase.

_____

Answers

16. It is the domain that manages program activities required to facilitate program definition, program delivery, and program closure.

17. Because programs have uncertainties, changes, complexities and interdependencies among the various components in them.

18. They depend on the type of program and tend to begin before funding is approved, or when the program manager is assigned.

19. 1. The program is authorized.
    2. The program roadmap is prepared to show expected results.
    3. The program business case is prepared.
    4. The program charter is prepared.
    5. The program management plan is prepared.

20. 1. The intended results of each component are produced.
    2. Individual components are authorized, planned, executed, transitioned, and closed.
    3. Benefits are delivered, transitioned, and sustained.

21. 1. The program benefits are transitioned to the sustaining organization.
    2. The program is formally closed.

Questions

22. List six activities that may be conducted by portfolio management before the program definition phase begins.

_____

23. What is the primary purpose of the program definition phase?

_____

24. What are the two sub-phases in program definition?

_____

25. When is the program manager selected?

_____

26. What is involved in program formulation?

_____

27. When is the program sponsor assigned?

_____

28. What are two key roles of the program sponsor?

_____

29. Why should the program manager be selected early in the program?

_____

Answers

22. 1. Develop concepts for products, services, or organizational outcomes.
    2. Determine scope frameworks.
    3. Develop initial requirements.
    4. Determine timelines.
    5. Determine deliverables.
    6. Determine acceptable cost guidelines.

---

23. To progressively elaborate the program's goals and objectives, define the program's outcomes and benefits, and seek program approval.

---

24. Program formulation and program planning.

---

25. During program formulation.

---

26. The development of the program's business case.

---

27. During program formulation.

---

28. To secure funding for the program and to select the program manager.

---

29. To guide the activities in program formulation and facilitate development of its required outputs.

Questions

30. List three activities in which the sponsoring organization, the sponsor, and the program manager work closely to prepare.

_____

31. List three reasons as to why studies of scope, resources, and cost are needed.

_____

32. What happens if the business case is developed before program formulation?

_____

33. List three reasons why an initial risk assessment is needed.

_____

34. What happens if the program is not authorized?

_____

35. When are the outputs of the program formulation phase updated?

_____

36. What is involved in program activities and integration management?

_____

37. How are program activities and integration management used throughout the program life cycle?

_____

Answers

30. 1. Studies and estimates of scope, resources, and cost.

    2. An initial risk assessment.

    3. A program charter and roadmap.

---

31. 1. To assess if the organization can deliver the program.

    2. To compare it with other initiatives to determine its priority in the program.

    3. To help create the business case if it was not done by the portfolio group.

---

32. It then is revisited and updated if needed.

---

33. 1. Analyze threats and opportunities.

    2. Determine the program's probability of successful delivery of benefits.

    3. Identify risk response strategies and plans.

---

34. The information is recorded in the lessons learned repository.

---

35. Throughout the program definition phase when business results are measured, and planned activities become more defined.

---

36. Optimizing or integrating the costs, activities, or effort of the components at the program level.

---

37. To use the resources, knowledge, and skills to deploy components.

Questions

38. List six decisions that are made in program activities and integration management.

_____

39. Why are program activities and integration management more cyclical and more iterative?

_____

40. What is all work performed in a program for program management considered?

_____

41. Why are program activities interdependent and complementary?

_____

42. How do program activities directly support individual components?

_____

43. What is the purpose of program integration management?

_____

44. What five activities in program integration management are needed to incorporate components into the program?

_____

Answers

38. 1. Competing demands and priorities.
    2. Risks.
    3. Resource allocations.
    4. Changes because of uncertainty and complexity.
    5. Interdependencies among components.
    6. Coordination of work.

---

39. Because adjustments are made based on outcomes and benefits if realignment is needed to strategic priorities.

---

40. Program activities.

---

41. The deliverables from one activity may be needed to produce those of another activity.

---

42. To ensure component activities help achieve program activities.

---

43. It is the core activity that occurs across the program life cycle.

---

44. 1. Identify
    2. Define
    3. Combine
    4. Unify
    5. Coordinate

Questions

45. Why is program infrastructure development performed?

_____

46. When does program infrastructure development begin?

_____

47. Can program infrastructure development be repeated?

_____

48. What are two purposes of program infrastructure development?

_____

49. When is the program core team formed?

_____

50. Who helps develop the program's infrastructure?

_____

51. How can the PMO support the infrastructure?

_____

52. What are three areas the PMO establishes for the program for consistency?

_____

53. What is another key element of the program's infrastructure?

_____

Answers

45. To investigate, plan, and evaluate the needed support system to help the program achieve its goals.

---

46. In the program definition phase.

---

47. At any time in the life cycle to update or modify the infrastructure.

---

48. To establish the management and resources needed for the program and the components.

---

49. In program infrastructure development.

---

50. The core team.

---

51. It may be part of it, but it supports management and coordination of the work of the program and its components.

---

52. 1. Policy
    2. Standards
    3. Training

---

53. The program management information system (PMIS).

Questions

54. What is the purpose of the PMIS?

_____

55. List eight items that can comprise the PMIS.

_____

56. Why are the program infrastructure resources separate and distinct from the resources managing the components?

_____

57. List four items included in program delivery management involving program components.

_____

58. List four items in the program delivery phase related to the activities performed.

_____

59. Who is responsible for initiating a request for a new component?

_____

60. Who approves the request to initiate a new component?

_____

Answers

54. To have tools to collect, integrate, and communicate information.

---

55. 1. Software tools.
    2. Documents, data, and knowledge repositories.
    3. Configuration management tools.
    4. Change management system.
    5. Risk data base and analysis system.
    6. Financial management systems.
    7. Earned value management activities and tools.
    8. Requirements management activities and tools.

---

56. Because the majority of the resources and program costs are managed at the component level.

---

57. 1. Management
    2. Oversight
    3. Integration
    4. Optimization

---

58. 1. Initiation
    2. Change
    3. Transition
    4. Closure

---

59. The program manager.

---

60. The program steering committee.

Questions

61. What is the program manager's role if the component is approved?

_____

62. Who determines if a component's initiation is accelerated or deferred?

_____

63. If the program manager has the authority to approve change requests, what happens?

_____

64. When does the program manager coordinate with a customer or sponsor to close or transition a component?

_____

65. What is the next step if the request to close or transition is approved?

_____

66. What happens if the component transition is approved?

_____

67. What do the transition and closure updates represent?

_____

68. What are three items that are affected with the approval or disapproval of a transition or closure request?

_____

69. What are three items involved in monitoring and controlling the program?

_____

70. Why is continuous monitoring important to the program manager?

_____

Answers

61. To determine if the priority of the existing components should be redefined.

62. It is defined by the program team and its needs.

63. They then are used in managing performance and determining if the program management plan requires changes.

64. When it reaches the end of its life cycle or as planned milestones are achieved.

65. A formal request is sent to the program steering committee for review and approval.

66. The program roadmap is updated.

67. A go/no-go decision and an approved change request.

68. 1. High-level milestones
    2. Scope
    3. Timing

69. Collecting, executing, and distributing performance information.

70. It provides insight into the program's health and identifies areas that require special attention.

Questions

71. How does monitoring support controlling?

_____

72. What are two examples of controlling activities?

_____

73. Who determines whether to execute requests for corrective or preventive actions?

_____

74. What happens if the requests exceed the program manager's authority level?

_____

75. What are two outputs from ongoing controlling?

_____

76. What is included in a program performance report?

_____

77. What two items are described in program performance reports?

_____

78. List six items that are part of these reports.

_____

Answers

71. It determines when controlling activities are needed to bring the program back into alignment with strategic priorities.

_____

72. Corrective or preventive activities.

_____

73. The program manager based on thresholds set by program governance on his or her authority level.

_____

74. The program steering committee makes the decision.

_____

75. Program performance reports and forecasts.

_____

76. A summary of the progress of all program components.

_____

77. Whether the program's goals will be met and its benefits delivered as planned.

_____

78. 1. Work accomplished
    2. Work to be completed
    3. Earned value
    4. Risks
    5. Issues
    6. Changes being considered

_____

Questions

79. How are forecasts used?

_____

80. What is the basis for the forecasts?

_____

81. When does component transition occur?

_____

82. Who is responsible for the stewardship of sustaining the benefits?

_____

83. When is the program closed?

_____

84. What are two examples of the internal and external conditions that may cause a program to be terminated?

_____

85. What happens if benefits continue to be realized after the program closes?

_____

86. What three activities are judged for a successful completion?

_____

87. What happens with the components before the program is closed?

_____

88. What happens with contracts before the program is closed?

_____

Answers

79. To assess whether achieved planned outcomes will occur and provide predictions of the program's future state.

80. Current information and available knowledge.

81. If there are immediate benefits or if the components deliver benefits at the same time.

82. It may be transitioned to another organization, entity, or subsequent program.

83. When the charter is fulfilled or if it needs to be terminated early given internal or external conditions.

84. Changes in the business case or if the expected benefits cannot be achieved.

85. They are managed as part of operations in the organization.

86. The approved business case, the actual outcomes of the program, and the organization's current and strategic objectives.

87. All components are completed or cancelled.

88. All contracts should be formally closed.

Questions

89. Who provides final acceptance to close the program?

_____

90. When is a final program report required?

_____

91. What critical information is included in the final report?

_____

92. List 10 items that may be part of the final report.

_____

93. When is knowledge transfer performed?

_____

94. What happens after knowledge transfer is complete?

_____

95. Why should lessons learned be readily accessible?

_____

Answers

89. The program steering committee.

90. As part of the program governance plan.

91. Information to improve the success of future programs.

92. 1. Assessments of finances and performance
    2. Lessons learned
    3. Successes and failures
    4. Areas identified for future improvement
    5. Risk management outcomes
    6. Unforeseen risks
    7. Customer sign off
    8. Reasons to close the program
    9. History of all program baselines
    10. An archive plan for documentation

93. When the program team assesses the program's performance and shares lessons learned.

94. The final report may require updates.

95. For existing or future programs for continuous learning and to avoid similar pitfalls.

Questions

96. How does knowledge transfer support benefit sustainment?

_____

97. What is the best way to release program resources when the program is closed?

_____

98. What is the best way to reassign resources at the component level?

_____

Answers

96. To provide the new supporting organization with relevant documentation, training, and materials.

---

97. To reallocate or reassign people and funding to other initiatives or programs.

---

98. To transition them to another component in the program or to another program in the organization.

Program Activities

Questions

1. List 10 program activities that support program governance and program management.

2. What is the purpose of the program supporting activities?

3. What type of coordination activities do program activities require?

4. List four high-level areas in program formulation that are assessed.

5. Why are areas assessed in program formulation?

6. Why are activities in program formulation exploratory?

Answers

1. 1. Program Change Management
   2. Program Communications Management
   3. Program Financial Management
   4. Program Information Management
   5. Program Procurement Management
   6. Program Quality Management
   7. Program Resource Management
   8. Program Risk Management
   9. Program Schedule Management
   10. Program Scope Management

---

2. To plan, monitor and control, and deliver program outputs and benefits.

---

3. Ones with functional groups in the organization but at a broader level than those involved with a single project.

---

4. 1. Scope
   2. Risks
   3. Costs
   4. Expected benefits

---

5. To confirm the program is a viable way forward for the organization and is aligned with strategic objectives.

---

6. They look at alternatives to ensure the best is aligned with strategy and organizational preferences.

Questions

7. In program formulation, are programs ever cancelled?

_____

8. What is the core program activity?

_____

9. What are the two activities from program governance?

_____

10. What is provided to program governance from program integration management and its supporting activities?

_____

Answers

7. Yes, if the activities show the program lacks a strong business case.

---

8. Program Integration Management.

---

9. Strategic guidance and the program charter.

---

10. The validity of the business case.

Initiating/Program Formulation

Questions

1. What is the primary document the program steering committee uses to authorize the program?

_____

2. What happens when the charter is approved?

_____

3. List 13 items in a program charter.

_____

4. What is meant by justification?

_____

5. What is meant by vision?

_____

6. List four items in assumptions and constraints.

_____

Answers

1. The program charter.

2. The program can start, and the program manager can apply resources to program activities.

3. 1. Justification
   2. Vision
   3. Strategic alignment
   4. Benefits
   5. Scope
   6. Benefit strategy
   7. Assumptions and constraints
   8. Components
   9. Risks and issues
   10. Timeline
   11. Resources needed
   12. Stakeholder considerations
   13. Program governance

4. Why the program is important and what it is to achieve.

5. The program's end state and how it will benefit the organization.

6. 1. Assumptions
   2. Constraints
   3. Dependencies
   4. External factors and how they affect the program's objectives.

Questions

7. What is involved in the components section of the charter?

_____

8. Why are risks and issues in the program charter?

_____

9. What is in the timeline in the program charter?

_____

10. What resource information should be included in the program charter?

_____

11. What are three items to include in the program charter on stakeholders?

_____

12. Why are stakeholder considerations in the charter useful?

_____

13. List three items to include in the charter on governance.

_____

14. When is the information in the charter on governance updated?

_____

15. What is the purpose of program change assessment?

_____

Answers

7. How the projects and other components are configured to deliver the program and its intended benefits

8. They are the initial risks and issues defined when the roadmap is prepared.

9. To show the length of the program and its key milestone dates.

10. The estimated costs and resource needs such as staff, training, and travel.

11. 1. The key stakeholders.
    2. The most important stakeholders.
    3. The initial strategy to engage them.

12. They help develop the communications management plan.

13. 1. The governance structure for the program
    2. The governance structure to guide and oversee the components.
    3. The program manager' authority.

14. In the program governance plan.

15. It identifies sources of change and helps develop the business case.

Questions

16. What are three sources of possible change?

_____

17. What is a goal of program change assessment?

_____

18. What are three outputs from program change assessment?

_____

19. Why is program communications management different from project communications management?

_____

20. Why is an initial assessment of program communications needs important?

_____

21. Why is it necessary to engage with the wide range of stakeholders and maintain communications with them?

_____

22. In program formulation, what can the program manager do to identify stakeholder expectations?

_____

23. In addition to the communications assessment, what are two other outputs of program communications assessment?

_____

24. What is the purpose of the initial cost estimate?

_____

Answers

16. 1. Enterprise environmental factors.
    2. Sensitivity of the proposed business case to changes in organizational strategy.
    3. Possible frequency and magnitude of changes.

17. To estimate the likelihood of the impact of the changes that could occur and determine how to respond to change proactively and not in a disruptive way.

18. 1. Input to the business case.
    2. Input to the program charter.
    3. Input to the program change management plan.

19. It affects a wider number of stakeholders with different communications needs, approaches, and methods of delivering information.

20. It is a key input to the charter.

21. To prevent more serious problems from occurring.

22. Conduct a survey to find out their expectations and interests in being involved in the program.

23. An input to the stakeholder engagement plan and program communications plan.

24. It is a part of the business case and assesses a level of confidence in the estimate.

Questions

25. What type of estimate is it?

_____

26. As a rough-order-of-magnitude estimate, what should be done?

_____

27. What are three outputs from Program Initial Cost Estimation?

_____

28. Why is a Program Information Assessment useful?

_____

29. Since the assessment is the output of this activity, what are the inputs from it?

_____

30. What is the purpose of the Program Procurement Assessment activity?

_____

31. What are two examples of programs that present unique procurement challenges?

_____

32. Why is this assessment prepared in the program definition phase?

_____

33. While the procurement assessment is the output of the Program Procurement Management Assessment, list the three areas where it is an input.

_____

Answers

25. A rough order of magnitude as there is limited information, time, and resources available.

26. Identify the nature and source of the costs that could not be estimated.

27. It is an input to the program business case, the program charter, and the detailed cost estimate in program planning.

28. The assessment shows the program's information needs to plan for financial or resource implications.

29. The business case, the program charter, and the information management plan.

30. It is a valuable input to the program charter.

31. Ones involving public-private partnerships or involving organizations or work in other countries.

32. To determine if there are special challenges or if procurement will represent significant work in program delivery.

33. 1. Business case
    2. Program charter
    3. Procurement management planning

Questions

34. What is involved in assessing program quality management?

35. What are two examples of quality constraints in program delivery?

36. In a quality assessment, what are important inputs to program costs and resources?

37. From the perspective of a quality assessment, what are considerations for quality and risk assessments?

38. While the output from the Program Quality Assessment is the assessment, this assessment serves as an input to three items – list them.

39. What are seven examples of resources required to plan and deliver a program?

40. Why should you should prepare an estimate of the resources for your business case?

41. The program resource requirements estimate is an output of which activity?

Answers

34. To determine the quality expectations, constraints, risks, and standards that should be evaluated.

---

35. Operational or regulatory standards.

---

36. Expectations of the quality of program outputs.

---

37. Ability of suppliers to comply with quality standards.

---

38. 1. Program business case.
    2. Program charter.
    3. Program quality management planning.

---

39. 1. People
    2. Office space
    3. Laboratories
    4. Data centers or other facilities
    5. Software
    6. Vehicles
    7. Office supplies

---

40. All required resources should be estimated and especially those with long-lead items.

---

41. Program Resource Requirements Estimation.

Questions

42. During the Program Initial Risks Assessment activity, what are two aspects of risk that should be estimated?

_____

43. What is risk appetite and how is it used in Program Initial Risk Assessment?

_____

44. While program risk assessment is the key output of the Program Initial Risk Assessment activity, it is an input to five documents; list them.

_____

45. What is the purpose of the Program Schedule Assessment activity?

_____

46. Where is the level of confidence in activity durations stated?

_____

47. Why are alternative activities considered in a program schedule assessment activity?

_____

48. While the schedule assessment is the output of the Program Schedule Assessment activities, it is an input to three other documents. List them.

_____

49. What is covered in a program scope management assessment?

_____

Answers

42. The key risks the program may encounter and the likelihood and impact of their occurrence.

---

43. Risk appetite is the organization's tolerance level to assess and deal with risks; it is helpful to understand the level of effort required to monitor and assess risks during program delivery.

---

44. 1. Program business case
    2. Initial cost estimate
    3. Program charter
    4. Program roadmap
    5. Program risk management plan

---

45. To evaluate expectations for delivery dates and benefit milestones.

---

46. As part of the program schedule assessment.

---

47. To identify whether there are alternatives that could be initiated if the schedule activities have excessive delays.

---

48. 1. Program charter
    2. Program roadmap
    3. Program schedule management plan

---

49. It includes boundaries, links to other programs or projects, and ongoing activities.

Questions

50. When is a program scope assessment required?

_____

51. How does the initial scope assessment support the scope statement?

_____

52. How is the input from the program scope assessment obtained?

_____

Answers

50. As part of the program charter and as an input to support initial cost, change, resources, risks, and schedule assessments.

---

51. It helps develop it from the program goals and objectives.

---

52. From the program sponsor or stakeholders through portfolio management or stakeholder alignment activities.

Planning/Program Planning

Questions

1. How is planning handled on a program?

2. When does program planning begin?

3. In program planning, list two activities that are used to develop the plan.

4. What is the program management plan?

5. How are these controls used?

6. What data are used to measure performance?

7. What is the purpose of performance measurement?

Answers

1. Programs use high-level plans that track component interdependencies and guide planning at the component level.

2. When the program steering committee approves the program charter.

3. 1. The initial program organization is defined.
   2. The program team is established.

4. The document used to integrate subsidiary plans and also to establish management controls and an overall plan to integrate and manage the program's components.

5. To measure performance against the plan.

6. Information is collected and consolidated from the components.

7. To ensure the program remains aligned with the organization's strategic objectives to deliver the promised benefits.

Questions

8. List five documents used to develop the project management plan.

_____

9. What is the key output in program planning?

_____

10. What happens if updates or revisions to the program management plan are needed?

_____

11. When is the program organization defined and the initial team deployed?

_____

12. How is the roadmap used in program planning?

_____

13. How are the management arrangements for program delivery used in the program plan?

_____

14. Why should the program plan be open for changes?

_____

Answers

8. 1. The organization's strategic plan.
   2. Business case.
   3. Program charter.
   4. Roadmap.
   5. Any other outputs from program formulation such as the outputs from assessments.

---

9. The program management plan.

---

10. They are processed through Governance.

---

11. In program planning.

---

12. It is included in the program plan.

---

13. They are included in the program plan to assist in monitoring and controlling.

---

14. It takes into consideration that the program's success is not measured according to performance against the baseline but is measured by benefit realization from program outcomes.

Questions

15. List the 12 supporting program activities.

_____

16. Why should a change management activity be established for the program?

_____

17. What seven items are part of the change management plan?

_____

18. What is a key focus of the change management plan?

_____

19. Who determines the level of change thresholds?

_____

20. What are the two outputs of the change management planning activity?

_____

Answers

15. 1. Program change management planning.
    2. Program communications management planning.
    3. Program cost estimation.
    4. Program financial framework establishment.
    5. Program financial management planning.
    6. Program information management planning.
    7. Program procurement management planning.
    8. Program quality management planning.
    9. Program resource management planning.
    10. Program risk management planning.
    11. Program scope management planning.
    12. Program scope management planning.

16. To administer changes the occur during the program.

17. 1. Establish change management practices and procedures.
    2. Capture requested changes.
    3. Evaluate each proposed change.
    4. Determine the disposition of each proposed change.
    5. Communicate the decision to impacted stakeholders.
    6. Document the change request.
    7. Authorize funding for the work

18. The impact of the change to the propose outcomes and the benefits expected.

19. The program steering committee.

20. The program change management plan and the level of change thresholds.

Questions

21. Why are communications so important on program?

_____

22. What is included in the program communications management planning activity?

_____

23. What is included in the program communications plan?

_____

24. Why is it important to clearly define communications requirements?

_____

25. Where should communications requirements to specific stakeholders be documented?

_____

26. When planning communications what are three areas to consider?

_____

27. What are two outputs from the communications planning activity?

_____

28. When is program cost estimating performed?

_____

29. What is meant by a tired funding process?

_____

Answers

21. Because of the time the program manager spends communicating with internal and external stakeholders about the program.

---

22. Timely and appropriate generation of information collection, distribution, storage, retrieval, and ultimate disposition.

---

23. It describes the how, when, and by whom information will be disseminated.

---

24. To facilitate information transfer from the program to the components and to the stakeholders with appropriate content ad delivery methods.

---

25. In the stakeholder register.

---

26. Stakeholder changes, cultural differences, and time zones.

---

27. The communications management plan and stakeholder requirements in the stakeholder register.

---

28. Throughout the course of the program.

---

29. One with a series of go/no-go decisions at each major program stage, and there is agreement to a financial plan and commitment to a budget only to the next stage.

Questions

30. How is a confidence factor applied to the cost estimate?

_____

31. What is an example of a statistical technique for confidence in the estimate?

_____

32. How is the confidence factor used?

_____

33. In addition to development and implementation costs, what else should be part of the program cost estimate?

_____

34. What comprises the total cost of ownership?

_____

35. How is the total cost of ownership used?

_____

36. What is another purpose of program cost estimates?

_____

37. What happens if the assumptions are not true in program delivery?

_____

38. How are the program cost estimates used by components?

_____

39. What are three outputs from program cost estimation?

_____

Answers

30. By using a weight or probability based on the program's risk and complexity.

---

31. Monte Carlo simulation.

---

32. To determine a range of potential costs.

---

33. Sustainment costs after the program is complete.

---

34. Full program life cycle costs including transition and sustainment costs.

---

35. To compare the expected benefits of one program to another to make a funding decision.

---

36. To identify any key assumptions used in the estimating process.

---

37. The program business case and program management plan may need to be reconsidered.

---

38. To guide component cost estimation.

---

39. Program cost estimates, program cost estimating assumptions, and component cost estimation guidelines.

Questions

40. What dictates the financial framework for the program?

_____

41. List five funding models.

_____

42. What is the impact on the funding timing?

_____

43. What is the objective in program financing?

_____

44. Why is the funding organization not a passive stakeholder on programs?

_____

45. What is the program's financial framework?

_____

46. What happens as the program's financial framework is developed and analyzed?

_____

47. What are the three outputs from program financial framework establishment?

_____

Answers

40. The type of program and the funding structure.

---

41. 1. Funded completely from one organization.
    2. Managed in one organization but funded separately.
    3. Funded and managed entirely from outside the program organization.
    4. Has both internal and external funding sources.
    5. Program funded by one or more sources, but components may be funded from other sources.

---

42. It affects the program's ability to perform.

---

43. To provide funding to bridge the gap between paying funds for development and realization of benefits.

---

44. Because of the large amount of money to fund programs.

---

45. A high-level plan to coordinate funding, determine constraints, and determine how funds are allocated.

---

46. Changes may impact the business case requiring its revision.

---

47. Program financial framework, business case updates, and updates to the communications management and stakeholder engagement plans.

Questions

48. What is involved in the Program Financial Management activity?

_____

49. What is documented in the program's financial management plan?

_____

50. How does the program's financial management plan expand on the financial framework?

_____

51. What else should be in the plan if the program is funded internally?

_____

52. What are three ways to consider if a program is funded internally?

_____

53. What is involved to develop the program's budget?

_____

54. Why should the budget be baselined?

_____

55. Why is it difficult to develop financial metrics?

_____

56. Who establishes and validates financial performance indicators?

_____

Answers

48. Identify financial sources and resources, integrate component budgets, develop the program's budget, and control program costs.

49. Funding schedules and milestones, initial budget, contract payments and schedules, financial reporting activities, and financial markets.

50. It describes managing risk reserves, potential cash flow problems, international exchange rate fluctuations, future interest rate changes, inflation, currency devaluation, financial local laws, trends in material costs, and contract incentive and penalty costs.

51. Scheduled contract payments, inflation, and other environmental factors.

52. Retained earnings, bank loans, or the sale of bonds.

53. Compile financial information and list income and payment details.

54. To use it to measure the program's finances.

55. Cause-and-effect relationships are difficult to determine to establish given the size and length of a program.

56. The program team and the program steering committee.

Questions

57. Why are financial measures important?

_____

58. Where should the program financial risks be incorporated?

_____

59. List the seven outputs from program financial planning.

_____

60. What is the purpose of the program information management plan?

_____

61. List five items that may be part of this plan.

_____

62. List four ways the information in this plan may be gathered and retrieved.

_____

63. When are program information distribution methods determined?

_____

Answers

57. Their results are used in making decisions to continue, cancel, or modify the program.

58. In the program risk register.

59. 1. Program financial management plan
    2. Initial program budget
    3. Program financial schedules
    4. Component payment schedules
    5. Program operational costs
    6. Inputs to the program risk register
    7. Program financial metrics

60. To describe how information assets will be prepared, collected, organized, and secured.

61. 1. Information management policies
    2. Distribution lists
    3. Appropriate tools
    4. Templates
    5. Reporting

62. 1. Manual filing systems
    2. Electronic systems
    3. Project management software
    4. Systems to access technical information

63. Once the program's information management system is determined.

Questions

64. What are the two outputs from the Program Information Management activity?

_____

65. What is program procurement management?

_____

66. What is the Program Procurement Management activity?

_____

67. What is the program procurement management plan?

_____

68. What are three useful techniques the program manager can use in preparing the procurement management plan?

_____

69. What contributes to successful program procurement management?

_____

70. As the program manager prepares the procurement management plan he or she determines four key items. List them.

_____

71. What is an example of the approach to competition?

_____

72. What is an example of external regulatory mandates?

_____

Answers

64. The program information management plan and the program information management tools and techniques.

65. Applying knowledge, skills, tools and techniques to acquire goods or services to meet the program's and the components needs.

66. It addresses activities needed to acquire goods and services and the unique procurement needs of the program and its components.

67. It describes how the program will acquire needed goods and services from outside of the performing organization.

68. Make-or-buy decisions, the PWBS, and available funding.

69. Early and intensive planning by the program manager.

70. 1. Whether a program-level procurement can meet some needs of components.
    2. The best type of contracts to use for the program and components.
    3. The best program-wide approach to competition.
    4. The best program-wide approach to balance external regulatory mandates.

71. The risk of a sole source competition versus that of a full and open competition.

72. The percentage needed to meet a small business mandate.

Questions

73. List three types of analysis that are useful in program procurement planning.

_____

74. What are three outputs from the Program Procurement Management Activity?

_____

75. What is the purpose of program quality planning?

_____

76. What does the program's quality plan describe?

_____

77. What do the program quality management activities determine?

_____

78. Why should the program manager prepare a quality policy?

_____

79. How is the business case used in quality management planning?

_____

80. Why is it beneficial to evaluate quality across the program?

_____

81. Why should a program quality manager participate in the planning activities?

_____

Answers

73. 1. Requests for Information
     2. Trade studies
     3. Market analysis

---

74. Program procurement standards, program procurement plan, and program budget plan updates.

---

75. It identifies relevant organizational or regulatory standards and how to satisfy them during the program.

---

76. How the organization's quality policies will be implemented.

---

77. The quality policies, objectives, and responsibilities needed to be successful.

---

78. To document the program's quality objectives and principles and share them with the component managers.

---

79. It includes the cost for the level of quality requirements, which then can be evaluated in quality management planning.

---

80. To combine quality tests and inspections to reduce costs if possible.

---

81. To verify quality activities and controls are applied and flow down to components including subcontractors.

Questions

82. While the program quality plan is the output of the Program Quality Management Planning activity, list six items it may contain.

83. Why is resource management at the program level different from that at the component level?

84. What is program resource management?

85. What are three examples of resources?

86. What is Program Resource Management Planning?

87. Why are the total component human resources less than the total quantity of resources needed for the program?

88. How should the program manager analyze resource availability?

89. What is useful to consider to determine the types of needed resources?

Answers

82. 1. Quality policy.
    2. Quality standards.
    3. Quality estimates of costs.
    4. Metrics, service level agreements, or memorandums of understanding.
    5. Checklists.
    6. Assurance and control specifications.

---

83. Because the program level works within bounds of uncertainty and balances the needs of the components.

---

84. It ensures all required resources are available for the component managers to enable program benefit delivery.

---

85. People, equipment, and materials.

---

86. Identifying existing resources and the need for additional resources.

---

87. The resources can be reallocated between components as a component is completed.

---

88. In terms of capacity and capability and how to allocate the resources across components.

---

89. Historical information.

Questions

90. What is the purpose of the resource management plan?

_____

91. What guidelines are in the program resource management plan?

_____

92. What happens if resources are not available in the program?

_____

93. What are the two outputs from Program Resource Management Planning?

_____

94. What is Program Risk Management Planning?

_____

95. What is the purpose of the program risk management plan?

_____

96. Describe three risk management planning activities.

_____

97. When should risk management planning be done?

_____

98. When should risk management planning be repeated?

_____

Answers

90. To forecast the expected level of resource use across components and relative to the master schedule to identify resource shortfalls or conflicts about scarce resources.

91. To make resource prioritization decisions and resolve resource conflicts.

92. The program manager works with the larger organization for assistance and may develop a statement of work to contract for resources.

93. Program resource requirements and the program resource management plan.

94. It identifies how to approach and conduct program risk management and that of the program's components.

95. It describes how risk management activities are to be structured and performed.

96. 1. Ensures the level, type, and visibility of risk management are supported.
    2. Identifies resources and time for risk management activities
    3. Establishes an agreed-upon basis to evaluate risks.

97. Early in the program definition phase.

98. Whenever there are major program activities.

Questions

99. What is the program's risk register?

_____

100. What is the most suitable way to manage program risks, adjust risk sensitivity, and monitor risks?

_____

101. What are two items that influence the program's risk management plan?

_____

102. How are risk profiles expressed?

_____

103. Why are the actions taken about risks important?

_____

104. What is an environmental factor to consider about program risks?

_____

105. What are two other factors that shape the risk management approach?

_____

106. List six predefined risk management approaches common in many organizations.

_____

107. What are two outputs from the Program Risk Management Planning activity?

_____

Answers

99. The document to record the results of risk analysis and planning.

---

100. Assess the organization's risk profile.

---

101. Risk thresholds and risk targets.

---

102. In policy statements and revealed in actions taken.

---

103. They how the organization's willingness to embrace high-threat situations or its reluctance to forgo high opportunities.

---

104. Market factors to the program and its components.

---

105. Culture and stakeholders.

---

106. 1. Risk categories
    2. Definitions of concepts and terms
    3. Risk statement formats
    4. Standard templates
    5. Roles and responsibilities
    6. Decision-making authority levels

---

107. Program risk management plan and program risk register.

Questions

108. What are four purposes of the Program Schedule Management Planning activity?

_____

109. How is the program schedule developed?

_____

110. What precedes the development of the program schedule?

_____

111. When is the initial program master schedule developed?

_____

112. How are the program delivery date and major milestones developed?

_____

113. When are the program's component milestones used in the program's master schedule?

_____

114. List three other activities in the program's master schedule.

_____

115. What three items are determined by the program's master schedule?

_____

Answers

108. 1. Determines the order and timing of the components to produce benefits.
     2. Estimates the time for each one.
     3. Identifies significant milestones.
     4. Documents the outcomes of each milestone.

---

109. Collaboratively with the schedules of the components.

---

110. The scope management plan and the PWBS.

---

111. Before that of the detailed schedules of the components.

---

112. By using the roadmap and the program charter.

---

113. They are included if they represent a program output or share an interdependency with other components.

---

114. 1. Stakeholder engagement activities.
     2. Program-level risk mitigation.
     3. Program-level reviews.

---

115. 1. Timing of individual components.
     2. When benefits will be delivered.
     3. Program external dependencies.

Questions

116. What happens once the program's master schedule is determined?

_____

117. What happens when a program is established based on existing components in schedule development?

_____

118. List three items essential to managing the program's master schedule.

_____

119. What is the purpose of the program's schedule management plan?

_____

120. What guidance should be included in the schedule management plan?

_____

121. What does the program schedule provide to stakeholders?

_____

122. What happens if the program schedule shows risks?

_____

123. How is the program roadmap used in the program schedule?

_____

Answers

116. The dates for the individual components schedule are identified.

117. The program's master schedule incorporates the milestones and deliverables from the individual component schedules.

118. 1. Maintain a logic-based network diagram.
     2. Monitor the critical path for component outputs with interdependencies.
     3. Focus on benefit realization of deliverables on the critical path.

119. To establish activities to develop, monitor, and control the schedule.

120. Coordinating changes to the schedule baselines and controlling activities across the components.

121. A visual representation on how the program will be delivered in its life cycle.

122. They should be included in the program's risk register.

123. It should be assessed periodically and updated so there is alignment; any roadmap change should be reflected in the schedule.

Questions

124. List four outputs from the Program Schedule Management activity.

_____

125. What is Program Scope Management Planning?

_____

126. What is the objective of Program Scope Management?

_____

127. What are three ways to describe program scope?

_____

128. What is a Program Work Breakdown Structure (PWBS)?

_____

129. What are three items in the PWBS?

_____

130. List five examples of artifacts.

_____

131. Where does decomposition of the PWBS stop?

_____

Answers

124. 1. Program schedule management plan.
   2. Program master schedule.
   3. Inputs to the program's risk register.
   4. Updates to the program roadmap.

---

125. It includes all activities in planning and aligning the scope with the program's goals and objectives.

---

126. To develop a detailed program scope statement with decomposition of the work into component program deliverables to deliver associated benefits.

---

127. Expected benefits and user stories or scenarios.

---

128. A deliverable-oriented hierarchical decomposition to show the total scope of the program.

---

129. Deliverables to be produced by components, artifacts, and PMO support deliverables.

---

130. 1. Plans
   2. Procedures
   3. Standards
   4. Processes
   5. Program management deliverables

---

131. At the level of control designated by the program manager; typically, the first one or two levels of a component.

Questions

132. List six uses of the PWBS.

_____

133. List four links from program-level deliverables

_____

134. What is a guideline to follow in developing the PWBS?

_____

135. When should a plan to manage, document, and communicate scope changes be defined?

_____

136. What is the purpose of the program scope management plan?

_____

137. What are three outputs from Program Scope Management Planning?

_____

Answers

132. 1. Framework to develop the program's master schedule.
    2. Defines management control points.
    3. Builds realistic schedules.
    4. Develops cost estimates.
    5. Helps organize work.
    6. Framework to report, track, and control work.

---

133. 1. Benefits.
    2. Stakeholder engagement activities.
    3. Program-level management.
    4. Component-level oversight.

---

134. To avoid decomposing component-level deliverables.

---

135. Once the scope is developed during the program definition phase.

---

136. To describe how the scope will be defined, developed, monitored, controlled, and verified.

---

137. Program scope statement, program scope management plan, and the PWBS.

Executing/Progam Delivery Management

Questions

1. When does the program delivery phase begin?

_____

2. What is included in the program delivery phase?

_____

3. Why is program delivery iterative and not linear?

_____

4. Where are the component management plans prepared?

_____

5. Since the program manager may not know all of the components needed in program definition, what are the next steps?

_____

6. List three phases for components in program delivery.

_____

Answers

1. When governance approves the program management plan.

2. Activities to produce intended results of the components according to the program management plan.

3. The component capabilities are integrated into the program to facilitate delivery of the program's benefits.

4. At the component level and then integrated into the program level.

5. To oversee components and re-plan or realign them to accommodate changes in program direction using adaptive change.

6. 1. Component authorization and planning.
   2. Component oversight and integration.
   3. Component transition and closure.

Questions

7. What is included in program delivery management?

_____

8. When does program delivery end?

_____

9. What is the purpose of component authorization?

_____

10. Where are these criteria found?

_____

11. When is component planning performed?

_____

12. What are four types of plans for a component?

_____

13. What happens in component status and integration?

_____

14. Can the program manager initiate a component to coordinate the work done by the other components?

_____

Answers

7. Management, oversight, integration, and optimization of the program components to deliver benefits so the organization realizes value.

8. When governance determines specific criteria for the phase are satisfied or if the program is terminated.

9. It involves initiating components based on specific organizational criteria and individual busies cases.

10. In the program's governance plan.

11. Throughout the program delivery phase.

12. 1. Project management plan
    2. Transition plan
    3. Operations plan
    4. Maintenance plan

13. Components provide status information, which ii integrated and coordinated into program activities.

14. Yes, it is useful for component deliverables and helps to coordinate benefit delivery.

Questions

15. What is component transition?

_____

16. List five examples of operational support activities.

_____

17. Where are the criteria to perform component transition activities and the organization's expectations defined?

_____

18. What happens before the program delivery phase ends?

_____

19. Who performs the final review of component transition and closure?

_____

20. When are program delivery management activities performed?

_____

21. Who typically requests initiating a component?

_____

22. What happens if a new component is approved?

_____

Answers

15. It addresses the need for ongoing activities from a component to an operational support function to achieve ongoing benefits.

16.   1. Product support
      2. Service management
      3. Change management
      4. User engagement
      5. Customer support

17. In the program governance plan.

18. Components are reviewed to verify benefits were delivered and to transition any remaining projects and sustainment activities.

19. The program sponsor and the program steering committee.

20. Throughout the program delivery phase as they relate to initiating, changing, transition, and closing of components.

21. The program manager.

22. The program manager redefines priorities of existing components to ensure optimal use of resources and manages interdependencies.

Questions

23. How are change requests handled?

_____

24. How is the program management plan affected by change requests?

_____

25. Who collaborates with the program manager to prepare a request to close or transition the program?

_____

26. How is the roadmap used in component transition?

_____

27. What two updates are reflected in the roadmap?

_____

28. What are reflected by these roadmap updates?

_____

29. What is included in the program delivery phase activities?

_____

30. What are seven activities in the program delivery phase?

_____

Answers

23. The program manager approves or rejects them if he or she has the authority to do so.

---

24. It may require changes if the change request is approved.

---

25. The customer or the sponsor.

---

26. It is updated.

---

27. Go/no-go decisions and approved change requests.

---

28. High-level milestones, scope, timing, of major stages scheduled during the program.

---

29. Activities required to coordinate and manage the program.

---

30. 1. Change control
    2. Reporting
    3. Information distribution
    4. Cost
    5. Performance
    6. Quality
    7. Risk

Questions

31. List the ten supporting processes in program delivery.

_____

32. What activities are part of Program Communications Management?

_____

33. List five types of people who may receive program communications.

_____

34. List eight outputs from Program Communications Management.

_____

Answers

31. 1. Program change monitoring and controlling
    2. Program communications management
    3. Program financial management
    4. Program information management
    5. Program procurement management
    6. Program quality assurance and control
    7. Program resource management
    8. Program risk monitoring and control
    9. Program schedule monitoring and control
    10. Program scope monitoring and control

32. The timely and appropriate generation, collection, distribution, storage, retrieval, and support of component communication to ensure alignment with the program's overall communications objectives.

33. 1. Clients
    2. Program sponsor
    3. Program steering committee
    4. Component manages
    5. Possibly the press and the public

34. 1. Status information
    2. Program change requests
    3. Program financial reports
    4. External filings with government and regulatory agencies
    5. Presentations for legislative bodies
    6. Public announcements
    7. Press releases
    8. Media interview and benefits updates

Questions

35. What is included in status information?

_____

36. What parts of the program provide status information?

_____

37. What is included in information about change requests?

_____

38. List five ways to distribute information.

_____

39. What are examples of electronic conferencing and communications tools?

_____

40. What are examples of program management tools?

_____

41. What is a social media tool?

_____

42. What are the primary methods to communicate day-to-day activities?

_____

43. What should be done regardless of the distribution method selected?

_____

Answers

35. Progress, cost information, risk analysis, and other information specific to a particular group

36. The overall program, projects, subsidiary programs, or other work

37. Notification about them to the program and component teams and the response to the request

38. 1. Face-to-face meetings and presentations.
    2. Electronic communications and conferencing tools.
    3. Electronic tools for program management.
    4. Social media, interviews, conference presentations, marketing, and publication articles.
    5. Internet, small group conversations, and staff meetings.

39. E-mail, fax, voicemail, telephone, video and web conferencing, and video presentations.

40. Web interfaces to schedules and project management software, meetings and virtual office support software, and collaborative work management tools.

41. Internet-focused group communications tools.

42. E-mails, small group conversations, and staff meetings.

43. It should remain in the program's control.

Questions

44. What happens if an incorrect message is sent to someone?

_____

45. Why is a full-time communications manager needed for some programs?

_____

46. What is included in the program's budget?

_____

47. What is the program measured against?

_____

48. What is the primary financial target?

_____

49. What are three cost items that need to occur before the program budget is baselined?

_____

50. What are two important parts of the budget?

_____

51. Where are the schedules and milestones identified when funding is received?

_____

52. What do the component payment schedules identify?

_____

53. What happens when the baseline is determined?

_____

Answers

44. It may cause problems for the program and lead to its termination.

---

45. Because communications are time consuming and challenging.

---

46. Costs for each component and resources to manage the program.

---

47. The baselined program budget.

---

48. The baselined program budget.

---

49. 1. Individual component costs.
    2. Contract costs.
    3. Program management and supporting activities costs.

---

50. Component payment schedules and program payment schedules.

---

51. In the program payment schedules.

---

52. How and when contractors are paid.

---

53. The program management plan is updated.

Questions

54. What are the three outputs of the Program Cost Budget activity?

_____

55. What are two examples of when initial cost estimates need to be updated?

_____

56. Why is it best to prepare a cost estimate as close to beginning the work?

_____

57. What happens if the component costs more than its estimate?

_____

58. When are cost estimates for components baselined?

_____

59. What happens if a contractor is responsible for a component?

_____

60. What is the output of Component Cost Estimation?

_____

61. What is a significant task in complicated or complex programs?

_____

62. Why is Program Information Management necessary for effective program management?

_____

Answers

54. 1. Updates to the budget baseline.
    2. Program payment schedules.
    3. Component payment schedules.

55. If components are not known when initial order-of-magnitude cost estimates are prepared or based on the current environment and cost considerations.

56. If the component cost is lower than planned, the program manager may present an opportunity to the sponsor for other products to be added later.

57. A change request is needed to determine if there is funding for corrective action.

58. When they are completed and then become the component's budget.

59. The cost is in the contract.

60. Component cost estimates.

61. Program Information Management.

62. Because of the extensive information exchanged among program management, component management, portfolio management, program stakeholders, and the organization's governance function.

Questions

63. What is involved in Program Information Management?

_____

64. How can errors or incorrect decisions be avoided by using information management?

_____

65. What is an invaluable aid to other program activities from information management?

_____

66. How is the information repository useful?

_____

67. What are two outputs from Program Information Management?

_____

68. What are lessons learned?

_____

69. Where are lessons learned knowledge acquired?

_____

70. List five areas in which lessons learned should be reviewed.

_____

Answers

63. It must be available to support communications management, requires archiving, and is a continuous task.

_____

64. By paying attention to accuracy and information timeliness.

_____

65. The program information repository.

_____

66. When there is a need to refer to past decisions or prepare trend analysis using historical information.

_____

67. Updates to the information repository and inputs to information distribution and program reporting.

_____

68. A compilation of knowledge gained.

_____

69. From similar past program or in data bases in the public domain.

_____

70. 1. Updates to the program stakeholder register.
    2. Updates to the program's risk register.
    3. Updates to the program's communications management plan.
    4. Consideration of changes to the program management plan.
    5. When new components are being considered.

_____

Questions

71. When is the lessons learned data base updated?

_____

72. What are the four outputs from the lessons learned data base?

_____

73. What is the key objective of conducting program-level procurements?

_____

74. What are four types of procurement standards?

_____

75. What is a common structure used by the program manager in Program Procurement Management?

_____

76. List the six outputs from Program Procurement Management.

_____

77. What happens once procurement standards are in place, and agreements are signed?

_____

Answers

71. When necessary and at the end of the program.

---

72. 1. Lessons learned reports.
    2. Inputs to updates for the stakeholder register and risk register.
    3. Inputs to updates to the communications management plan updates.
    4. Input to changes in the program management plan.

---

73. To set standards for components.

---

74. 1. Qualified seller lists.
    2. Pre-negotiated contracts.
    3. Blanket purchase agreements.
    4. Formalized proposal evaluation criteria.

---

75. To have all procurements centralized and conducted at the program level.

---

76. 1. Request for quote.
    2. Request for proposal.
    3. Invitation to bid.
    4. Proposal evaluation criteria.
    5. Agreement administration plan.
    6. Signed agreements.

---

77. Administration of contracts and their closeout are transitioned to components.

Questions

78. What happens at the component level in contract administration?

_____

79. What happens when contracts are administered at the program level?

_____

80. Why is the program manager visible in procurements?

_____

81. What are three outputs from Program Contract Administration?

_____

82. What happens in program delivery in Program Resource Management?

_____

83. What is the program manager's role in resource prioritization?

_____

84. How is human resource planning involved?

_____

85. Why are resource fluctuations common in program delivery?

_____

86. How does the program manager balance program needs with available resources?

_____

Answers

78. Component managers report results and closeouts to the program manager.

79. Component managers report deliverable results, contract changes, and contract issues to the program team.

80. To ensure the budget is spent properly to obtain benefits.

81. Performance/earned value reports, ongoing progress reports, and vendor/contract performance reports.

82. The program manager monitors, controls, and adapts resources for benefits delivery.

83. To prioritize scarce but needed resources and optimize resource use.

84. In identifying, documenting, and assigning program roles and responsibilities.

85. The need for staff, facilities, equipment and other resources may change often because of supply and demand.

86. By working with component managers with available resources.

Questions

87. How should resource prioritization decisions be made?

---

88. What happens when there are changes to existing components or when new components begin?

---

89. What are two outputs from Program Resource Management?

Answers

87. Based on guidelines in the program management plan.

---

88. The program resource management plan may need changes.

---

89. Program resource management decisions and updates to the program resource management plan.

Controlling/Program Performance Monitoring and Controlling

Questions

1. How do program managers handle monitoring?

2. What is the purpose of Program Change Monitoring and Controlling?

3. What monitoring factors could create changes to the program?

4. What is a program change request?

5. Where are program change requests recorded?

6. Why are program change requests analyzed?

Answers

1. They track component progress to ensure overall goals, schedule, budget, and benefits are met.

2. Activities in which modifications to documents, deliverables, or activities in the program are identified, approved, or rejected.

3. Factors internal or external to the program that show a need for change.

4. A formal proposal to modify any program document, deliverable, or baseline.

5. In a change log.

6. To determine their urgency and impact on program baseline documents or program components.

Questions

7. What happens if there are multiple ways to implement the change?

_____

8. Who makes the decision concerning a change request?

_____

9. List three items that need to be done once the change request decision is made.

_____

10. How are the results communicated to stakeholders?

_____

11. What are the two outputs from Program Change Monitoring and Controlling?

_____

12. Why is program reporting an element of program communications?

_____

13. What is involved in the Program Reporting activity?

_____

14. What is included in Program Reporting?

_____

15. How is status and deliverable information provided to stakeholders?

_____

Answers

7. The costs, risks, and other options are assessed to select the approach that is best for benefit delivery.

8. The program manager or the program steering committee.

9. 1. Record it in the change log.
   2. Communicate the result to stakeholders.
   3. Update component plans as needed.

10. By following the communications management plan.

11. Approved change requests and updates to the program change log.

12. It supports program governance and stakeholder engagement.

13. It consolidates performance and reporting data to enable stakeholders to learn how resources are being used to deliver benefits.

14. Information across projects, subsidiary programs, and program activities.

15. Through the information distribution activity.

Questions

16. What type of information is provided to team members and components?

_____

17. What happens with information provided by customers?

_____

18. What are three outputs from Program Reporting?

_____

19. What is part of the required reports to sponsors or program agreements?

_____

20. When is Program Financial Management a monitoring and controlling activity?

_____

21. Why is it critical to monitor program finances and control budget expenditures?

_____

22. What happens if the program's costs exceed its planned budget?

_____

23. What happens if there are minor budget overruns?

_____

Answers

16. General and background information about the program.

___

17. The program team gathers it, analyzes it, and distributes it back within the program.

___

18. Required reports to program sponsors or program agreements, customer feedback reports, and periodic reports and presentations.

___

19. Formats and reporting frequency.

___

20. When the program receives initial funding and begins paying expenses.

___

21. It ensues the program meets the goals of the financing unit or of the higher organization.

___

22. It may not justify its business case and may be subject to termination.

___

23. The program may require an audit and management oversight and should be justified.

___

Questions

24. List nine typical financial management activities.

_____

25. What are two earned value indices that are used?

_____

26. Why are contract expenditures monitored?

_____

27. What is involved when changes are communicated to governance and auditors?

_____

28. Why are program infrastructure costs managed?

_____

29. When are contract payments made?

_____

30. When is the budget updated and rebaselined?

_____

31. When are individual component budgets closed?

_____

Answers

24. 1. Identify actions that cause budget baseline changes.
    2. Monitor the environment for financial impacts.
    3. Manage changes when they occur.
    4. Monitor cost reallocation impacts and results among components.
    5. Monitor contract expenditures.
    6. Implement earned value.
    7. Identify impacts to components from overruns or underruns.
    8. Communicate changes to governance and to auditors.
    9. Manage program infrastructure costs.

---

25. Schedule performance index and cost performance index.

---

26. To ensure funds are distributed as stated in the contract.

---

27. They include changes at the program and component levels.

---

28. To ensure they are within established parameters.

---

29. As specified in the contract, within the financial infrastructure, and with the status of deliverables.

---

30. Throughout the program when changes are approved with significant cost impacts.

---

31. When the component work is completed.

Questions

32. How are new financial forecasts communicated?

_____

33. How often are new financial forecasts prepared?

_____

34. When are approved program and component financial changes incorporated into the appropriate budget?

_____

35. List six outputs from Program Financial Management.

_____

36. What is the purpose of Resource Interdependency Management?

_____

37. How does the program manager perform the interdependency responsibility?

_____

38. What happens if resources are no longer required?

_____

39. What happens to ensure the program resource management plan accounts for change?

_____

40. What is the output from Resource Interdependency Management?

_____

Answers

32. According to the program's communications management plan.

33. On a regular basis.

34. As they are approved.

35. 1. Contract payments
    2. Closed component budgets
    3. Program budget baseline updates
    4. Approved change requests
    5. Program management plan updates
    6. Corrective actions

36. The program manager shares resources among different components such that interdependencies do not delay benefits delivery.

37. By controlling the schedule for scarce resources.

38. The program manager releases them to other programs.

39. The program manager works with component managers.

40. Updates to the program resource management plan.

Questions

41. In terms of Program Risk Monitoring and Controlling, list the three key areas.

_____

42. What are four reasons why risk monitoring is conducted?

_____

43. How does program risk monitoring and control affect the components?

_____

44. What three activities are involved in Program Risk Identification?

_____

45. List 11 people who may participate in risk identification.

_____

46. Why is risk management iterative?

_____

Answers

41. 1. Program risk identification

    2. Program risk analysis

    3. Program risk response development

---

42. 1. To see if program assumptions are still valid.

    2. To see if the assessed risks have changed and any trends that have occurred.

    3. To see if program risk priorities and procedures are being followed.

    4. To see if cost or schedule reserves require modifications.

---

43. They require coordination with component risk monitoring and controlling.

---

44. 1. Determining risks that could affect the program.

    2. Documenting risk characteristics.

    3. Preparing for successful risk management.

---

45. 1. Program manager

    2. Program sponsor

    3. Program team members

    4. Risk management team

    5. External subject matter experts

    6. Customers

    7. End users

    8. Component manager

    9. Stakeholders

    10. Risk management experts

    11. External reviewers

---

46. New risks may evolve and become known.

Questions

47. Why should the format of risk statements be consistent?

_____

48. What is the importance of risk identification?

_____

49. What is the output from Program Risk Identification?

_____

50. What is Program Risk Analysis?

_____

51. How can Program Risk Analysis provide program benefits?

_____

52. What risk techniques are used to support program management decisions?

_____

53. Why is the information from qualitative and quantitative risk analysis important?

_____

54. What other useful Information is provided concerning these reserves?

_____

55. Why is information about the impact of negative and positive possible risks important?

_____

56. What is a difference between the impact of possible risks at the component level and that at the program level?

_____

Answers

47. To compare program risk events.

48. To provide information so the risk can be analyzed and prioritized.

49. Updates to the program's risk register.

50. It integrates relevant program component risks.

51. By managing the interdependencies of component risks.

52. Qualitative and quantitative risk analysis techniques.

53. It provides information about the amount of contingency and management reserve to be set aside if risks occur.

54. Assessments about costs, schedules, and performance outcomes for components and their interdependencies.

55. It can affect delivery of program benefits.

56. The time scale as component risks may occur in the short term and program risks in the long term.

Questions

57. What are the three outputs from Program Risk Analysis?

_____

58. What is Program Risk Response Management?

_____

59. Is the program risk contingency the same as that at the component level?

_____

60. Based on the program manager's direction, list six items of the program risk register that may be updated.

_____

61. List the six outputs from Program Risk Response Management.

_____

62. What is Program Schedule Monitoring and Controlling?

_____

63. What is involved in Schedule Monitoring and Controlling?

_____

Answers

57. 1. Proposed risk responses.
    2. Updates to the program's risk register.
    3. Periodic risk reports showing tends in threats and opportunities.

---

58. The actions the program manager uses to reduce negative risk consequences or to enable realization of potential benefits.

---

59. No – it is not a substitute for component risk contingency.

---

60. 1. Specific actions to take with the selected response strategy.
    2. Budget and schedule activities to implement the response.
    3. Contingency plans and trigger conditions that lead to their execution.
    4. Fallback plans to a risk that occurred if the primary response is inadequate.
    5. Residual risks that may remain after the response has been implemented or deliberately accepted.
    6. Secondary risks that occur after implementing a risk response.

---

61. 1. Direction to implement the risk response.
    2. Program risk register updates.
    3. Contingency and management reserves.
    4. Change requests.

---

62. Ensuring the program produces the required capabilities and benefits on time.

---

63. Tracking and monitoring the start and finish of high-level program activities and milestones against the program master schedule.

Questions

64. What is needed to maintain an accurate and current master schedule?

_____

65. What is needed for successful program management?

_____

66. What is involved in schedule control?

_____

67. Why should the program master schedule be reviewed?

_____

68. When is it necessary to accelerate or decelerate components?

_____

69. Why is it important to identify early deliveries?

_____

70. When is it necessary to approve deviations from the components' schedules?

_____

71. Why is it necessary to update the master schedule regularly?

_____

72. When are new components added or removed?

_____

Answers

64. Updating the master schedule and directing changes to component schedules.

65. Alignment of program scope, cost, and schedule.

66. Identifying both slippages and opportunities to accelerate the program schedule and for risk management.

67. To determine the impact of component schedule changes on other components and on the program.

68. To help achieve program goals.

69. To provide opportunities to accelerate the program.

70. To realize benefits if there are component performance deviations.

71. Because of the length of the program and its complexity and as components end and others begin.

72. Based on approved change requests to meet program goals.

Questions

73. What document is revised if there is a major change to the program master schedule?

_____

74. What are the three outputs from Program Schedule Monitoring and Controlling?

_____

75. What is the purpose of Program Scope Monitoring and Controlling?

_____

76. What are four sources of scope changes that may impact the program significantly?

_____

77. How should scope monitoring and controlling be done?

_____

78. What are five activities to perform in scope monitoring and controlling?

_____

79. What happens if major scope changes are approved?

_____

Answers

73. The program roadmap.

74. Updates to the program master schedule, the program risk register, and the roadmap.

75. To ensure successful program completion.

76. 1. Stakeholders
    2. Program components
    3. Unidentified requirements issues
    4. External sources

77. By following the change management and scope management plans.

78. 1. Capture requested scope changes.
    2. Evaluate requested scope changes.
    3. Determine the disposition of the change request.
    4. Communicate the decision to involved stakeholders.
    5. Record the change request and supporting detail

79. Updates to the scope statement and program management plan may be needed.

Questions

80. Who is responsible for determining affected components when a scope change is requested and for updating the PWBS?

81. What is the program manager's role if there are scope changes in a large program and the number of components affected is difficult to determine?

82. List the four outputs from Program Scope Monitoring and Controlling.

Answers

80. The program manager.

_____

81. Manage scope to the allocated level of components and avoid controlling scope that is further decomposed by component manager.

_____

82. 1. Update the program scope statement.
    2. Disposition of scope change requests and document why the decision was made.
    3. Update the program management plan.
    4. Update the PWBS.

_____

Closing/Program Closure

Questions

1. Who approves program closure?

2. What does the steering committee use to determine whether to close a program?

3. Provide two reasons for termination.

4. How do program and project managers work together during closing?

5. What are three activities that occur during program closure?

6. What happens in the program closure phase?

7. Who is consulted to see if the program met all desired benefits?

Answers

1. The program steering committee.

2. It closes the program when is benefits are realized, its objectives are realized, or the program should be terminated.

3. 1. There is a change in organizational direction.
   2. The planned benefits may not be achievable.

4. To ensure project outputs and outcomes are effectively transitioned to the program and project benefits are assimilated and sustained.

5. The program is transitioned and closed, the program is terminated early, or the program is transferred to another program.

6. All program activities are completed to transition benefits to the sustaining organization and to close the program in a controlled way.

7. The program steering committee.

Questions

8. Who is consulted to see if another program or sustaining activity will oversee the program's benefits?

_____

9. List four items that are done if a sustaining activity is needed.

_____

10. What happens when the transition activities are complete?

_____

11. When does program closure begin?

_____

12. What are two goals of program closure during this phase?

_____

13. When does benefit sustainment occur?

_____

14. What is the role of the program charter in closeout?

_____

15. What are examples of internal or external conditions in closure?

_____

16. What happens with benefits during closure?

_____

Answers

8. The program steering committee.

---

9. 1. Transition resources
   2. Transition responsibilities
   3. Transition knowledge
   4. Transition lessons learned

---

10. The program manager receives approval from the sponsoring organization to close the program.

---

11. When the components have delivered all of their outputs, and the program has started to deliver its benefits.

---

12. To release resources and support the transition of remaining assets.

---

13. It transcends the program components as it is performed as the program is closed.

---

14. The program is closed when the charter is fulfilled, or if internal or external events bring the program to an early end.

---

15. Thy may include changes in the business case that no longer make the program necessary, or they may serve to determine that the program's benefits cannot be realized.

---

16. They may have been realized or may continue to be realized by operations in the organization.

Questions

17. What are three ways used to determine if the program has been successful?

---

18. What are three things to do before the program is officially closed?

---

19. Before the program steering committee closes the program, a final report may be needed. List ten items to include in it.

---

20. When is knowledge transfer performed?

---

21. Should the final report be updated?

---

22. What happens to the program's human resources?

Answers

17. 1. Approved business case
    2. Actual program outcomes
    3. Goals and strategic goals of the organization

---

18. 1. Complete components
    2. Cancel components
    3. Formally close contracts

---

19. 1. Financial and performance assessments
    2. Lessons learned
    3. Successes and failures
    4. Identified areas for improvement
    5. Risk management outcomes
    6. Unforeseen risks
    7. Customer sign off
    8. Reasons to close the program
    9. History of all baselines
    10. Program archive plan for documentation

---

20. When the program is complete, the program team assesses the program's performance and shares lessons learned.

---

21. Yes, after the knowledge transfer is complete.

---

22. They are released as the program is closed; being reassigned to other initiatives, programs, or components needing the same skill sets.

Questions

23. What happens if extra funds remain?

_____

24. What is the definition of the program closure phase?

_____

25. What are three possible outcomes as the program is closed?

_____

26. What two activities are performed by the program steering committee during program closure?

_____

27. What happens if the program work is transitioned to another program or sustaining activity to oversee the benefits this program was to deliver?

_____

28. What happens after the transition activities are completed?

_____

29. List five supporting activities in program closure.

_____

Answers

23. They are reallocated to other programs, components, or other initiatives.

---

24. The activities necessary to transition the program's benefits to the sustaining organization and formally close the program.

---

25. 1. It is transitioned and closed.
    2. It is terminated early.
    3. Work is transitioned to another program.

---

26. 1. Determine if the program has met all of its desired benefits and transition activities are complete, including component transition.
    2. Determine if there is another program or sustaining activity that can oversee the ongoing benefits the program was to deliver.

---

27. Work may be needed to transition the resources, responsibilities, knowledge, and lessons learned to the other program.

---

28. The program manager received approval from the sponsoring organization to formally close the program.

---

29. 1. Program Financial Closure
    2. Program Information Archiving and Transition
    3. Program Procurement Closure
    4. Program Resource Transition
    5. Program Risk Management Transition

Questions

30. What happens with costs in Program Financial Closure?

_____

31. What are residual activities required to oversee the ongoing benefits?

_____

32. What are three ways to structure these residual activities?

_____

33. In Program Financial Closure, what are three items that occur as the program nears completion?

_____

34. How are the final financial reports communicated?

_____

35. List three items that show Program Financial Closure is complete.

_____

36. List six outputs from Program Financial Closure.

_____

Answers

30. Estimate may be needed for the costs to sustain the benefits the program created.

31. Activities that remain after costs are captured in operations, maintenance, or other activities initiated in the program delivery phase as components are completed.

32. 1. They may be structured as an individual project.
    2. They may be structured as a new program.
    3. They may be incorporated as new work in a separate portfolio or program in a new or an existing operations function.

33. 1. The program budget is closed.
    2. Final financial reports are communicated.
    3. Unspent monies are returned to the funding organization.

34. As described in the program communications management plan.

35. 1. Sustainment budgets are developed.
    2. Benefits are delivered.
    3. Sustainment has started.

36. 1. Input to the final program report.
    2. Updates to the program financial management plan.
    3. Inputs to the knowledge repository.
    4. Documentation of any new tools and techniques used into the knowledge management system.
    5. Financial closing statements.
    6. Closed program budget.

Questions

37. What happens in Program Information Archiving and Transition?

_____

38. What may be in the scope of Information Archiving and Transition?

_____

39. How is the transfer of program knowledge used to support tosupport benefit sustainment?

_____

40. How is a final lesson learned report prepared?

_____

41. How is the lessons learned report used?

_____

42. What are the two outputs from Program Information Archiving and Transition?

_____

43. List three Program Procurement Management activities.

_____

44. What happens if a program is closed early?

_____

Answers

37. Collect program records and organize them for use in the organization often because of legal reasons or to support other activities or other programs.

___

38. Collecting and archiving records and documents from program components.

___

39. It provides documentation, training, or materials.

___

40. The program manager assesses program performance and collects observations from team members.

___

41. To inform the governance and management of other programs in the organization, so they can avoid any pitfalls encountered by this program in program delivery.

___

42. Updates to organizational assets and the lessons learned report to organizational governance boards.

___

43. 1. All deliverables have been completed satisfactorily.
    2. All payments have been made.
    3. No outstanding contractual issues remain.

___

44. The contracts are terminated early to avoid unnecessary costs.

___

Questions

45. List three outputs from Program Procurement Closure.

_____

46. What happens in Program Resource Transition?

_____

47. What happens in component reassignment?

_____

48. What is the role of program governance in releasing resources?

_____

49. What is the output from Program Resource Transition?

_____

50. What is involved in Program Risk Management Transition?

_____

51. How is the appropriate organizational risk register used?

_____

52. Who manages these risks?

_____

53. What is the output of Program Risk Management Transition?

_____

Answers

45. 1. Contract closeout reports
    2. Lessons learned updates
    3. Closed contracts

---

46. Team members and funding are reallocated or reassigned to other initiatives or programs.

---

47. Resources are transitioned to another component in execution that requires a similar skill set.

---

48. It releases resources as part of activities that lead to program closure.

---

49. Resources released to other parts of the organization.

---

50. To determine if there are any remaining risks that could jeopardize benefits realization by the organization.

---

51. Remaining risks and supporting analysis and response information are transferred to this register.

---

52. A different organization group responsible for benefit realization or the PMO.

---

53. Inputs to other organizational risk registers.
